FULL CIRCLES

By bringing together gender, women's experience of space and life course, *Full Circles* offers a novel perspective to all three. It shows how historical, cultural and economic contexts, changing fertility patterns, class position and mobility, state policies and personal motivation influence the geographies of women's lives, constraining choices and providing options.

Full Circles describes the differences and commonalities in women's lives and expectations in post-industrial and developing countries from childhood to old age. Exploring the changing terrains of women's experience over the life course, the book shows how conditions in one stage affect opportunities in another. It highlights the importance of cohort, gender ideologies and the contradictory nature of women's roles in the home, the workplace and the community, and examines how these are reflected spatially.

Throughout, the aim is to reveal the critical intersections between space and time, place and person in women's lives, and the larger social relations with which they are mutually determined.

Cindi Katz is Assistant Professor of Environmental Psychology at the Graduate School and University Center of the City University of New York. **Janice Monk** is Executive Director of the Southwest Institute for Research on Women at the University of Arizona.

INTERNATIONAL STUDIES OF
WOMEN AND PLACE
Edited by Janet Momsen, *University of California at Davis*
and Janice Monk, *University of Arizona*

The Routledge series of *International Studies of Women and Place* describes the diversity and complexity of women's experience around the world, working across different geographies to explore the processes which underlie the construction of gender and the life-worlds of women.

Other titles in this series:

DIFFERENT PLACES, DIFFERENT VOICES
Gender and development in Africa, Asia and Latin America
Edited by Janet H. Momsen and Vivian Kinnaird

'VIVA'
Women and popular protest in Latin America
Edited by Sarah A. Radcliffe and Sallie Westwood

FULL CIRCLES

Geographies of women over the life course

Edited by
Cindi Katz and Janice Monk

London and New York

First published 1993
by Routledge
11 New Fetter Lane, London EC4P 4EE

Simultaneously published in the USA and Canada
by Routledge
29 West 35th Street, New York, NY 10001

Typeset in Baskerville by J&L Composition Ltd, Filey, North Yorkshire
Printed and bound in Great Britain by
Mackays of Chatham PLC, Chatham, Kent

British Library Cataloguing in Publication Data
A catalogue record for this title is available from the British Library.

Library of Congress Cataloging in Publication Data
Full circles: geographies of women over the life course / edited by
Cindi Katz and Janice Monk.
p.   cm.—(International studies of women and place)
Includes bibliographical references and index.
ISBN 0–415–07552–1.—ISBN 0–415–07562–9 (pbk)
1. Women—Social conditions—Cross-cultural studies.   2. Women—
Employment—Cross-cultural studies.   3. Life cycle, Human—Cross-
cultural studies.   4. Human geography.   I. Katz, Cindi, 1954–
II. Monk, Janice J.   III. Series.
HQ1154.F86   1993
305.4—dc20
92–19678
CIP

CONTENTS

List of plates   vii
List of figures   viii
List of tables   ix
List of contributors   xi
Acknowledgements   xiii

1 WHEN IN THE WORLD ARE WOMEN?   1
Janice Monk and Cindi Katz

2 WOMEN AND WORK ACROSS THE LIFE COURSE:
MOVING BEYOND ESSENTIALISM   27
Geraldine Pratt and Susan Hanson

3 ELIMINATING THE JOURNEY TO WORK: HOME-
BASED WORK ACROSS THE LIFE COURSE OF
WOMEN IN THE UNITED STATES   55
Kathleen Christensen

4 GROWING GIRLS/CLOSING CIRCLES: LIMITS ON
THE SPACES OF KNOWING IN RURAL SUDAN AND
US CITIES   88
Cindi Katz

5 'HE WON'T LET SHE STRETCH SHE FOOT':
GENDER RELATIONS IN TRADITIONAL WEST
INDIAN HOUSEYARDS   107
Lydia M. Pulsipher

6 WOMEN, WORK AND THE LIFE COURSE IN THE
RURAL CARIBBEAN   122
Janet Momsen

7 GENDER AND THE LIFE COURSE ON THE
FRONTIERS OF SETTLEMENT IN COLOMBIA   138
Janet G. Townsend

CONTENTS

8 OLD TIES: WOMEN, WORK AND AGEING IN A
COAL-MINING COMMUNITY IN WEST VIRGINIA   156
Patricia Sachs

9 LIFE COURSE AND SPACE: DUAL CAREERS AND
RESIDENTIAL MOBILITY AMONG UPPER-MIDDLE-
CLASS FAMILIES IN THE ÎLE-DE-FRANCE REGION   171
Jeanne Fagnani

10 LOCAL CHILDCARE STRATEGIES IN MONTRÉAL,
QUÉBEC: THE MEDIATIONS OF STATE POLICIES,
CLASS AND ETHNICITY IN THE LIFE COURSES
OF FAMILIES WITH YOUNG CHILDREN   188
Damaris Rose

11 WOMEN'S TRAVEL PATTERNS AT VARIOUS
STAGES OF THEIR LIVES   208
Sandra Rosenbloom

12 WOMEN, THE STATE AND THE LIFE COURSE
IN URBAN AUSTRALIA   243
Ruth Fincher

13 MAKING CONNECTIONS: SPACE, PLACE AND THE
LIFE COURSE   264
Cindi Katz and Janice Monk

Notes   279
References   289
Index   310

PLATES

4.1 Children's life in the *hosh* 91
4.2 Family within their houseyard 92
5.1 Typical houseyard in Montserrat 109
5.2 Elderly woman resident of houseyard in Montserrat 115
7.1 El Distrito: a girl of fourteen about to leave the region 149
8.1 On the sidewalk facing the coal route 160
8.2 A young girl in front of her house in the late 1940s 161

FIGURES

1.1   Global distribution of the female population aged 0–14 years   7
1.2   Global distribution of the female population aged 15–64 years   11
1.3   Global distribution of the female population aged 65 years and over   14
5.1   Location of Montserrat   111
6.1   World regional economic activity patterns of men and women by age group   125
6.2   Barbados female age-specific worker rates, 1946–80   126
6.3   Distribution of farmers by age, Nevis, 1979   132
6.4   Distribution of farmers by age, Barbados, 1987   132
9.1   Île-de-France region   175
10.1   Major influences on parents' selection of childcare mode for young school-aged children   203
11.1   Percentage of parents who link trips to work   214
11.2   Percentage of parents who link trips from work   215
11.3   Percentage of parents who make trips solely for their children, by age of children   217
11.4   Frequency of trips made solely for children under six   219
11.5   Children's most frequent travel mode as reported individually by parents of children under six   221
11.6   Children's most frequent travel mode as reported individually by parents for children aged six to twelve   222
11.7   Trips made solely for Dutch children of different ages by their parents   224
11.8a   Frequency of trips made solely for Dutch children by their parents; children under six   225
11.8b   Frequency of trips made solely for Dutch children by their mothers; children aged six to twelve   225
11.9   Licence holding by age and sex, 1988   228
11.10   Average annual miles driven by males and females of various ages   233
11.11   1983 per capita vehicle miles travelled for selected trip purposes, by sex, for people over sixty-five   234

TABLES

2.1 Gender differences in industries of employment   32

2.2 The gender composition of the occupation types   32

2.3 Family circumstances of employed women   34

2.4 Percentage of employed women in each occupational type, by marital status   36

2.5 Employers' opinions about the marital status of female and male employees   38

2.6 Household responsibilities of married women employed in different occupational types   40

2.7 Opportunities for part-time and shift work in manufacturing and producer services firms in four local areas of Worcester   41

2.8 Married women's evaluation of job attributes in general   44

2.9 Household responsibilities of married women   47

6.1 General worker rates in Barbados, 1891–1980   124

6.2 Economic activity rates for men and women in 1980   127

6.3 Occupations of women by age, Montserrat, 1980   128

6.4 Types of employment of male and female small-scale farmers in Barbados, Nevis and St Vincent   129

6.5 Gender differences in types of employment of small-scale farmers, Montserrat   129

6.6 Sources of income in small-farm households, by gender   131

6.7 Seasonal differences in weekly hours worked by male and female operators of small farms in Barbados and Nevis   134

7.1 Population structures by gender and age in each survey community   141

7.2 Female : male ratios by age group in each survey community of those who 'sometimes' or 'always' participate in tasks   144

7.3 Female : male ratios of adults (aged over twelve) in each survey community who 'always participate' in tasks   150

8.1 National origins of West Virginia miners   158

TABLES

9.1 Some housing and demographic characteristics in each area of the Île-de-France region 174

10.1 Conceptualizing modes of childcare provision: dimensions of analysis applied to the Québec case 190

10.2 State intervention in childcare provision in the Province of Québec 192

11.1 Percentage of Dutch children travelling alone, and most frequent travel mode 226

11.2 1983 per capita vehicle trips annually by retired people living with someone, and living alone 229

11.3 1983 per capita trips by vehicle, urban and rural areas 231

11.4 Male and female travel behaviour by age, 1983 235

11.5 Time consumed by alternative modes for average one-way trip now taken as vehicle driver for people over sixty-five 236

12.1 Aged care services provided by or affiliated with selected Melbourne local councils, 1988 257

CONTRIBUTORS

Kathleen Christensen is Professor at the Environmental Psychology Program, Graduate School and University Center of the City University of New York.

Jeanne Fagnani is Researcher at the Centre Nationale de la Recherche Scientifique and Scientific Advisor, Caisse Nationale des Allocations Familiales, Paris.

Ruth Fincher is Senior Lecturer at the Department of Geography, University of Melbourne.

Susan Hanson is Professor at the Graduate School of Geography, Clark University, Worcester, MA.

Cindi Katz is Assistant Professor of Environmental Psychology at the Graduate School and University Center of the City University of New York.

Janet Momsen is Professor at the Department of Geography, University of California at Davis.

Janice Monk is Executive Director of the Southwest Institute for Research on Women at the University of Arizona.

Geraldine Pratt is Associate Professor at the Department of Geography, University of British Columbia.

Lydia M. Pulsipher is Associate Professor at the Department of Geography, University of Tennessee.

Damaris Rose is Associate Professor at INRS-Urbanisation, Université du Québec.

CONTRIBUTORS

Sandra Rosenbloom is Director of the Drachman Institute, University of Arizona.

Patricia Sachs is Director of Social Solutions, Mount Vernon, New York.

Janet G. Townsend is Lecturer at the Department of Geography, University of Durham.

ACKNOWLEDGEMENTS

In editing this book we received valuable research assistance from Harouna Ba, Eric Graig and Penny Waterstone; Sho-ling Lee brought her skills in non-mechanized cartography to the preparation of maps for Chapters 1 and 9; Mary Contreras displayed great patience in compiling the manuscript in its final form. Neil Smith and Sallie Marston offered thoughtful commentary on several chapters. Pat Mora, whose poetry always captures women's experiences and their sense of place, graciously allowed us to quote *Stubborn Woman*. We appreciate the support of all these people who made our work possible.

1

WHEN IN THE WORLD ARE WOMEN?

Janice Monk and Cindi Katz

About 1860, when she was in her mid-thirties, my great-grandmother sailed from England to Australia with her husband and four or five children, despite her mother's apprehensions that this was a dangerous move. She had three more children in Sydney – the last, my grandmother, when she was forty. Child-bearing and -rearing occupied a great portion of her life. What else she did I do not know, but taking regular care of elderly parents was not part of her life's experience.

My grandmother, born in 1865, spent most of her life in one area of Sydney. She married at about the age of twenty-five and had four children, never working outside the home. Her husband's income was always small and social services were limited. At the end of his working life, faced with financial and health problems and the onset of the 1930s Depression, she and my grandfather came to live with my recently married parents and their baby son. She remained in this extended family till she died at the age of eighty-one. Born when she was seventy-two, I experienced her as an important companion of my childhood, but as an old woman who virtually never left the house and whom we did not leave there alone for more than a few hours.

My mother, born in 1901, has had yet a different life course and geography. Leaving school at the age of fourteen, she became a 'comptometrist' with a firm of accountants, working until her marriage at twenty-eight. She had two children – a son when she was twenty-nine and a daughter when she was thirty-five. Caring simultaneously for her parents, husband, and children, weathering hard times in the Depression, and living on her husband's modest income till she was widowed at fifty-nine, her spatial world was mainly limited to the neighbourhood and to those places she could reach on foot or by public transport. Then, as a widow, with a state pension and benefits, some support from her children, and years of experience as a careful money manager, she was financially able as well as motivated to take some vacations, visiting relatives outside Sydney and me in the United States, thus widening her world and acquiring new experiences for later memory. At the age of eighty-eight, no longer able to live alone, she moved to a 'hostel' in an outer suburb of Sydney, partially subsidized by church and state agencies, where, with other women (and a few men) mostly in their eighties and

1

nineties, she receives assistance in daily living. Her space is primarily one room, her view mostly open land. She feels isolated.

Like my great-grandmother, I moved across the world, though as a single woman, at the age of twenty-four. Unlike my forebears, I obtained considerable education, the beneficiary of a variety of state scholarships and institutional support from the age of eleven. Like them, I married in my late twenties. Unlike them, I had no children. Also unlike them, I support myself financially, now live alone, and travel widely nationally and internationally for work and pleasure. Unlike my great-grandmother, I am able to visit my family almost every year. But, living in the United States, I cannot anticipate the significant state support for my old age which my mother has received, nor the family care that was my grandmother's.

As I work on this book, reflection on the life experiences of the women in my family brings home to me how generation and historical context, changing fertility patterns, class position and class mobility, state policies and personal motivations intersect with the geographies of women's lives, constraining choices and providing options. As a daughter, I regret that my mother, at ninety, does not have the company and family care her mother experienced. As a middle-aged woman, I value the chances I have had and the choices I have made to create a different geography and to try to share with others some aspects of women's collective experience.

J.M.

My own reflections on the life course cross space rather than time, revealing my 'having to take into account the simultaneity and extension of events and possibilities' (Berger 1974: 40, as cited in Soja 1989: 22). As I contemplate my relationships with and to the women in rural Sudan with whom I lived for a year while conducting fieldwork on children's learning and knowledge, the socially constructed nature of each aspect of the life course becomes more vivid.

When I went to Howa I was nearly twenty-seven years old and in a long-term relationship with a man. Settling in a couple of months before my husband arrived I lived with Leila, a sixteen-year-old school-leaver from the city who had just come to Howa to teach reading, writing and basic home science to the women. I was her first student. Despite having had my own household for almost ten years, my skills were next to useless in the face of charcoal stoves, dirt floors, distant water supplies and handfuls of goat meat. I was grateful to learn how to be an efficient and clean homemaker from this teenager living away from her mother for the first time.

At first I felt an affinity with other young married women, thinking that we were at the same stage in our lives except that they had young children. I soon realized that most of them were nineteen or twenty years old. Women of my own age already had upwards of four children, some as old as ten or eleven. Becoming aware of these disjunctures in our lives, I understood why everyone in Howa found it inordinately amusing that someone would postpone childbearing to

2

complete a university degree and, given my advancing years, called upon me repeatedly to explain my lack of children. They elicited and listened to my explanation with a mixture of amusement and pity, which did not stop them from questioning me closely on other occasions about birth control.

In recent years as I have written about my work in Sudan (and entered my mid-thirties) it has been unnerving to realize that one of the women I considered a mother to me was all of thirty-seven years of age. At the same age as I (finally) began to feel like a 'grown-up' – degree completed, professional position secured, seriously considering having children – my fictive 'mother' had completed her family – married to the family patriarch, her youngest child sixteen; already a grandmother. She has made it clear to me that the family would prefer me to return with a baby than with more books. (I suspect my real mother might be in on this particular plot as well.)

These intersections point at once to the ties between biological life stages and their social construction. They also suggest the latitude in timing (and spacing) of life course stages, depending upon access to particular resources, and their interdependence on the choices and practices of those with whom we share our lives. These issues and questions connect production and reproduction, crossing public and private spheres of material social practice.

Clearly my life choices were in part framed by my access to educational resources and facilitated by both reproductive freedom and the health and economic independence of my parents. My particular choices have certainly attenuated aspects of the life stage associated with young adulthood. Yet the organization of and articulation between production and reproduction in the United States remain rooted in a sociospatial and political-economic structure that is not supportive of combining work in both spheres. Elements of the constraints of this structure came home to me repeatedly in Sudan, as I witnessed both the extended family in operation and the closer convergence of the spaces of production and reproduction which freed parents from full-time child care and supported their participation in other activities.

I witnessed this most poignantly in an occurrence that reveals again some of the cross-currents in the life course as constituted in different social, cultural and economic settings. Towards the end of my stay in Sudan, Medina, one of the women closest to me, had a stroke and died, pregnant at thirty-four. The family was, of course, devastated by her death, as was I, but what made it remarkable was the way the extended family structure sustained her immediate family, which included seven children from two to seventeen years of age, through the loss. This helped fill in the yawning gaps of her absence, even for her youngest children, who had their father, sisters, brothers, grandparents, aunts, uncles and cousins in the same household compound. A similar loss in the United States is generally much more destabilizing to the family, requiring, for example, family members to move house, enormous investments in family time at the expense of outside work activities, and/or great financial outlays.

When I returned to Howa a few years later, Medina's oldest daughter, by that

3

time fifteen, was nine months pregnant. When she went into labour, all the women of her family were with her during an arduously long night that ended in a stillbirth. I, twice her age, still childless, in some ways her 'fictive mother', sat at her side and thought of Medina.

These reflections bring home the diversity in women's experiences across space, time, class and culture, but also distil some of the structural similarities on which they pivot. Theorizing across these geographical settings may enable us to identify and examine some of the underlying processes within and against which women construct their lives.

C.K.

As Western feminist scholarship began to mature in the 1980s, attention turned increasingly to the significance of context in shaping women's lives and to the intersection of gender with other forms of difference, especially race, ethnicity and class. Reminded by women in other parts of the world that their experiences and visions are not encompassed by Western models, and as geographers who believe in the importance of place in people's lives, we were struck by international data assembled in the mid-1980s which revealed that one-third of the world's female population is under fifteen years of age and that this proportion reaches 45 per cent in sub-Saharan Africa (Sivard 1985: Table 1). Likewise, we were reminded that female life expectancy varies widely geographically – the difference between the highest and lowest countries is a startling forty-seven years (Sivard 1985: 24). Interest in the geographic meanings of these demographic variations, together with our own research on children (Katz) and older women (Monk), stimulated us to try to go beyond the body of feminist work in geography (and many other fields) which largely addresses the lives of women in their middle years, especially those given over to child-bearing and -rearing, and which generally keeps separate the experiences of women in First World and Third World contexts. We therefore assembled a group of scholars to describe and interpret the geographies of women's lives in an array of settings from the perspective of the life course, hoping that in this way we could extend understanding of the diversity and commonalities of women's experiences and of the roles of space and place in shaping those experiences.

In this introduction we will first present some basic demography, then review concepts that we have found useful in illuminating women's changing geographies – their uses of space, their relationships to place, and the ways in which place and space constrain and offer opportunities over women's life course. Finally, we will briefly introduce the essays that follow.

THE AGES OF WOMEN: CONTEMPORARY WORLD PATTERNS

Though Shakespeare's 'Seven Ages of Man' has been widely quoted over the centuries since its writing, his categories do not speak to women's lives, nor to the historical and cross-cultural diversity of experiences. Indeed, defining life in terms of a set of 'stages', especially if these are linked to chronological ages, is fraught with difficulty, especially if we adopt a comparative perspective. Comanches, for example, have been reported as identifying five stages, the Kikuyu six for males and eight for females, the Andaman Islanders twenty-three for men and the Incas ten (Falk *et al.*, cited in Chaney 1990: 43). If we recognize that a woman's life expectancy at birth is as low as 32–3 years in Kampuchea, Sierra Leone and Afghanistan but over seventy-eight years in France, Japan and Switzerland, we can hardly assume that 'stages' will be closely linked to chronological age or be experienced in universal ways. Yet we generally ignore these differences in research on women's lives, assuming in some fundamental way that 'a mother is a mother is a mother'.

Chronological age by itself does not define the roles and statuses a woman may have, the work she may do, or the timing of marriage, child-bearing or attainment of varying degrees of power in the family and society. Nevertheless, some understanding of the demography of women, country by country, does help us to identify how certain of their activities, such as caring for children or the elderly, will vary in salience across geographic contexts. Thus we will review the distribution of women around the world on the basis of age, using three broad categories: 0–14 years, 15–64 years and sixty-five years and over. We chose these categories mainly because that is how the data are available, but we also consider them of some consequence, despite the qualifications we have already expressed.

The contrast in the population profiles of rich and poor countries is well known. Underdeveloped nations, almost without exception, are characterized by a triangular profile, the wide base indicating many more young than old people; while wealthy industrial countries, with similar numbers in each age cohort, have more symmetrical and narrow silhouettes, signifying enduring patterns of low fertility. The first form signals not only higher dependency ratios, but portends continued population growth as growing numbers reach fertility. The latter profile connotes stabilized population growth and more of a balance between dependants and caretakers, although growing numbers of older people mark a new form of dependence, with dramatic life course implications, as Ruth Fincher discusses in her chapter in this book. Figures 1.1, 1.2 and 1.3 map these differences. Given our concern with the life course of women, we want to highlight some of the gendered short- and long-term consequences of these distributions.

We emphasize that it is not population size which creates problems; rather that it is the global and national distribution of wealth, power and other resources which makes any particular demographic situation problematic or not. The global recession and stupendous levels of Third World debt, for example, have stymied virtually all advances in education and health in these countries. Under circumstances of growing impoverishment and debt, many nations of Africa, Asia, South America and the Caribbean must work harder just to stay in place, given both the overall numbers of children and youth, and high rates of population growth.

There are sobering indications that they are failing. For example, since the Second World War the numbers of six- to eleven-year-olds not in school in the underdeveloped countries declined steadily, reaching a plateau in 1986 at 50 million, since when they have increased significantly for the first time in four decades, to 60 million (UNICEF 1990). Likewise, in the least developed countries (characterized by UNICEF as those with the highest rates of under-five mortality), the percentage of those enrolled in first grade who completed primary school dropped from 49 per cent in 1980–4, to 40 per cent in 1985–7. Among the next cluster of countries – largely in Africa and Asia, with a small number in South America – there was a decline from 70 to 64 per cent over the same periods. These trends, which point to economic difficulties at both the national and household levels, mean, of course, that as we reach the end of the millennium, the number of adult illiterates will grow and overall literacy rates will rise more slowly throughout much of the world.

As advances in education and health are slowed or reversed, the consequences for females and thereby society as a whole are critical. The very countries with the highest numbers of young people have the fewest resources – human, financial, infrastructural – for providing education or improving health. There are no easy conclusions or comparisons, however. Two brief examples, amplified below, illustrate this point. First, the availability of education does not correlate with whether or not girls are enrolled at similar levels to boys. Some countries with low levels of enrolment reach equal rates for both sexes, others with high levels of enrolment educate the majority of boys but few girls. The reasons for these differences reflect the intersection of political economy, cultural ideologies concerning education, specific sociospatial patterns and cultural ecologies and the gender division of labour at all scales. These interrelationships, which connect global with local processes, are dynamic and context-specific. Some of their determinants and consequences for particular sites and populations are explored in the chapters by Janet Townsend and Cindi Katz. Second, evidence is mounting that in the face of scarce resources, the preference for male children is exacting an enormous toll in female infanticide, abortion

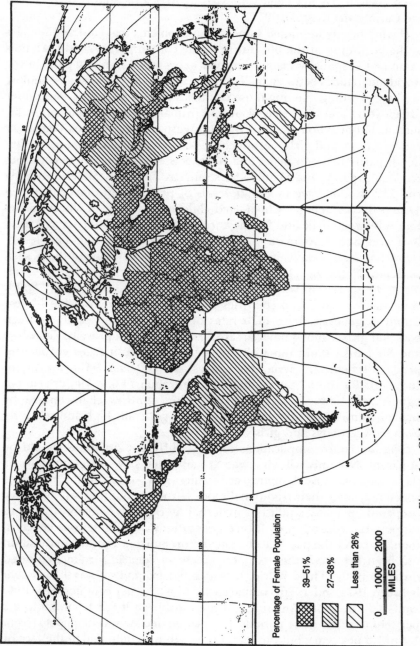

Figure 1.1 Global distribution of the female population aged 0–14 years

and neglect that results in death. However, these practices appear to be largely limited to Asia, where they are often exacerbated by strict government controls on family size. Despite devastating poverty, this phenomenon has not been witnessed in Africa or South America and the Caribbean (Kristof 1991).

Having briefly identified some of the problems that confront females at the beginning of the life course in the contemporary world, we now examine the spatial variations in their expression. In many parts of the world inequalities between males and females in their access to schooling are long-standing. Among those countries with high populations under fifteen years and low levels of overall primary school enrolment, some, such as Guinea, Somalia and Burkina Faso, exhibit serious disparities between boys' and girls' enrolment. Elsewhere, enrolments are low but the disparity between males and females is less; among these countries are Mali, Uganda and Haiti. Others, such as Tanzania, Malawi, Rwanda and El Salvador have moderate levels of primary enrolment (between 47 and 65 per cent of school-aged children) with no significant differences between male and female enrolment. Several countries in Africa, Asia and South America with large numbers of young people are characterized by relatively high levels of enrolment. Some of these, such as Benin, Nepal and Togo, exhibit disparities in boys' and girls' enrolment. This pattern, which favours boys, does not occur in any South American or Caribbean countries for which data are available – these countries generally have high rates of enrolment and parity between boys and girls. Among other areas where this pattern prevails, Zimbabwe and Sri Lanka stand out with 100 per cent of both sexes enrolled in primary school. In Syria the figures are 100 and 94 per cent, in Indonesia 99 and 97 per cent and in Nicaragua 74 and 79 per cent (all figures for 1986–8, UNICEF 1990). In countries characterized by smaller youth ratios, enrolment figures are generally high and relatively even between boys and girls.

The life course implications of these different situations should be apparent. Not only will girls who are educated have different skills, as they confront fast-changing economies and sociocultural situations which may alter their traditional modes of work, but they will be better prepared, as many studies of maternal literacy have demonstrated, to support the education of future generations. It is of serious concern, then, that over the past decade, long before universal primary education has been achieved, the rate of increase in female enrolment slowed throughout the Third World. In Africa, for example, it was 9.1 per cent between 1975 and 1980, compared with just 2.7 per cent between 1980 and 1988; in South America and the Caribbean it fell from 3.9 to 2.2 per cent over the same periods. In Asia the decline was negligible (from 1.2 to 1.0 per cent), but the rates of growth were minimal in the face of conditions in which 43 per cent of the total enrolment was female (UNESCO 1990). These reversals in progress are illustrative of the historically and geographically specific mutual determinations between the global political economy, development economics, cultural values and the division of labour. The particular constellation of these factors in any location will have a substantial bearing upon life course decisions, as well as upon the articulation of these with larger sociocultural and political-economic structures.

It remains the case that in much of the Third World, especially in the poorest countries of Africa and Asia, only a small minority of children attend secondary school. Just a fraction of these are female. As we discuss in more detail below, many women in these countries marry and begin bearing children in their teenage years. This pattern exacerbates extant health problems, because infant health and survival are compromised both by early pregnancy and with each birth from the same mother. Low birthweight babies (two kilograms or less) are more common among teenage mothers, for example. A study in the United States revealed that 13.8 per cent of the babies born to mothers under age fifteen had low birthweights, whereas among mothers of fifteen to nineteen years of age, 9.3 per cent were low birthweight; and among mothers of twenty to twenty-four years of age, the figure dropped to 7 per cent (Population Reference Bureau 1989). A study from Brazil, one of the countries with moderate numbers of births to mothers younger than fifteen years of age, revealed infant mortality rates of 124 per 1,000 live births among mothers under eighteen years of age. For mothers between eighteen and twenty-four years old, the figure drops by a third. While the same study found an increase in infant mortality with each new birth, the increase was substantial at the seventh birth and after. Similarly, the shorter the interval between births, the higher the infant mortality; under two years the figure is 138, while with a four-year interval, it drops to 50 per 1,000 live births (UNICEF 1990). These figures reveal some of the ways in which socially constructed life-course patterns, such as age of marriage and child-bearing, have serious implications for health and education in general.

Available statistics on infant mortality reveal, in the main, slightly higher rates for boys than girls in almost all countries. Likewise, female life expectancy at birth exceeds males' almost everywhere. Available data on malnutrition are not reported by sex, but evidence from many parts of the world indicates that males continue to be favoured in the distribution of resources, whether this be food, healthcare or maternal attention. When resources are scarce, we may expect girls to suffer more than boys. Recent demographic reports suggest widespread and substantial 'excess female mortality'. Girls are being aborted, killed and neglected at staggering rates in several regions, and these practices are showing up increasingly in skewed population ratios. In India and China alone, conservative estimates point to over 52 million missing females (that is, the number of females that should be alive according to expected population ratios and rates of population growth). A Swedish study suggested that in countries without strong patterns of discrimination, about 130 infant boys die for every 100 infant girls. In China, however, only about 112 boys die for every 100 girls (Johansson and Nygren 1991; Kristof 1991). Here it is calculated that among infants under a year old, 44,000 female deaths annually result from unequal treatment (Kristof 1991). Other Asian countries are experiencing this phenomenon as well, but it remains uncommon in Africa, South America and the Caribbean. The preference for boys, which is already having a pronounced effect on the sex ratios in much of Asia, has profound implications for the future. Not only will ratios of males to females be much higher, thus altering marriage, sexual and child-bearing patterns, but given that throughout the world most of the 'caring work' is done by females, it remains to be seen who will look after everyone. This trend illustrates, in an extreme and disturbing way, how the work of production and reproduction over the life course is profoundly altered by everyday material social practices and how these are articulated with larger social relations of production and reproduction at all scales.

In virtually all countries the years between fifteen and sixty-four include those in which women carry their maximum responsibilities for productive and reproductive work, generating income and caring for partners, children and the elderly. The proportion of the female population available for such work varies considerably among countries, however (Figure 1.2). The percentages are highest in eastern and western Europe, the United States and Canada, Australia and New Zealand, China and Japan; they are lowest in most African countries, parts of the Middle East and Central America. Intermediate values occur in much of South America and the Caribbean and south and south-east Asia.

To some extent, these differences can be linked to the nature of economic development and, indeed, lead one to ask how development is related to the availability of women for work. Yet this distinction is simplistic, for the distribution also reflects differences in fertility, child and maternal mortality and women's longevity which are shaped by cultural values as well as political-economic structures and socio-economic practices, reflected for example in the quality of healthcare afforded women or the demand for household labour. Perhaps the most important implication of the distribution is an obvious one: the countries with the highest proportions of children (Figure 1.1) are, by definition, those with the lowest proportions of women available for productive and reproductive work. Even though children may assist with work, this

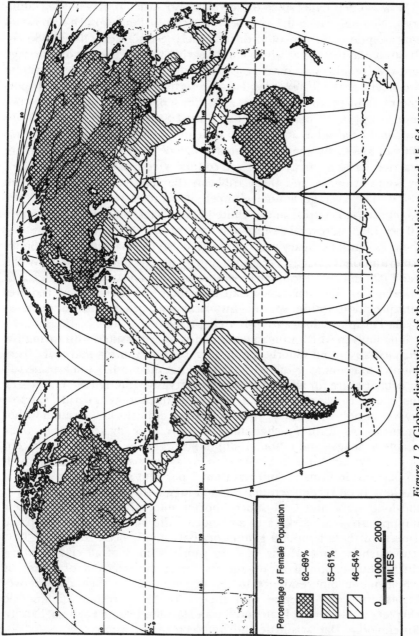

Figure 1.2 Global distribution of the female population aged 15–64 years

circumstance surely contributes to the long hours of work and small amount of time available for leisure, sleep and other physiological needs that have been documented for women in some African and Asian countries (Seager and Olson 1986: 13).

How women experience these years also varies widely among countries. Not only does the number of children (and women's access to contraception and abortion) differ widely (Seager and Olson 1986: 7–9), so too do the age at which child-bearing begins, the age of marriage and the frequency of marriage and divorce. A few examples illustrate the extent of the differences. Data on fertility indicate that the proportion of all births to teenagers is as low as 1 per cent in Japan and 4 per cent in Egypt, 13 per cent in the United States and Thailand, 20 per cent in El Salvador and 29 per cent in Cuba (Population Reference Bureau 1989). By age twenty, an average of 50 per cent of women are married across an array of African countries, compared with 38 per cent in South America and the Caribbean (Population Reference Bureau 1989). The average annual number of divorces reaches over 20 per 1,000 married couples in the United States but stands at between 5 and 10 per 1,000 in several eastern and western European countries, among them Finland, Hungary and (formerly West) Germany and some South American ones such as Venezuela and Uruguay. Rates fall to below 2.5 per 1,000 in Costa Rica, Ecuador, Italy and Sri Lanka. What patterns of marriage or non-marriage mean in women's lives, however, requires interpretation within specific historical and geographic contexts, as Lydia Pulsipher illustrates in her chapter.

The ways in which women arrange for or are assisted with caring for their children and elderly family members also differ markedly from country to country. While expectations that the mother will assume full or primary care for all dependants are pervasive (though see Pulsipher), the extent to which the State and other institutions provide support varies not only by country, but also by class, ethnicity and location within countries, as Ruth Fincher and Damaris Rose make clear in their chapters. Seager and Olson (1986) indicate that as many as 37 per cent of pre-school children in Sweden are in day-care centres, compared with 15 per cent in Canada, and less than 1 per cent in Nicaragua. While it has long been recognized that women juggle their schedules both over the long term and over much shorter intervals, to cope with their multiple roles as workers in and out of the labour force, the spatial aspects of these temporal gymnastics have rarely been addressed. Our studies show some of the ways in which women of all classes adjust their spatio-temporal arrangements, from the selection of places of employment and residence to the organization of their daily movements, in order to cope with their care-taking responsibilities (see chapters by Geraldine Pratt and Susan Hanson, Jeanne Fagnani, Sandra Rosenbloom, Damaris Rose and Kathleen Christensen).

Women's participation in productive work also varies widely geographically. Global patterns of women's labour are well summarized by Momsen and Townsend (1987). The inadequacies of international statistical data on women's work are widely known (see, for example, Beneria 1982), especially as they fail to deal with unpaid and informal sector activities and part-time employment. Within the limitations of these data, however, Seager and Olson have identified an array of countries, especially in Scandinavia and eastern Europe, where more than 65 per cent of women aged fifteen and over are working for wages or trade; and others, especially in the Middle East, where their documented labour force participation falls below 10 per cent (1986: 16). Age of entry into the workforce and withdrawal from it, however, also vary considerably by geographic context, as Janet Henshall Momsen, Janet Townsend, Geraldine Pratt and Susan Hanson and Kathleen Christensen reveal in their chapters in this book.

The geographic distribution of older women (Figure 1.3) around the world in the early 1980s presented a simpler pattern than that of the younger age groups. Countries clearly have either high or low proportions of women over age sixty-five. Globally, these women account for seven per cent of the female population, but proportions range from 2 or 3 per cent in most countries of sub-Saharan Africa to 16–19 per cent in most of western Europe (Sivard 1985: Table 1).

By the twenty-first century, however, the distribution of older women will be somewhat different because of current patterns of fertility and changes in life expectancy world-wide. Although most African countries will continue to have low proportions of older women, this group will increase in importance in South and Central America, the Caribbean and in China. Estimates for South and Central America and the Caribbean region as a whole are that women who turn sixty in the year 2000 can expect, on the average, to live an additional twenty years (Anstee 1990: 3). Already the proportions of older women in the populations of Argentina, Barbados and Uruguay are comparable to those found in the United States, Canada and Japan. Although in several Latin American countries – for example, Bolivia, Peru, Venezuela and the Dominican Republic – the proportions of women over age sixty-five in the early 1980s were similar to those in African nations, fertility is generally lower (except in Bolivia), and in many cases, life expectancy already more than ten years higher (Sivard 1985: Table 1). Of course, to understand changes in population structure we also have to examine immigration trends, especially in the Caribbean.

Unfortunately, we know relatively little about the experiences of older women around the world, even in the European Community, where one-third of the female population is over fifty years of age, 12 million women are over age seventy-five and 2.5 million are over age eighty-five

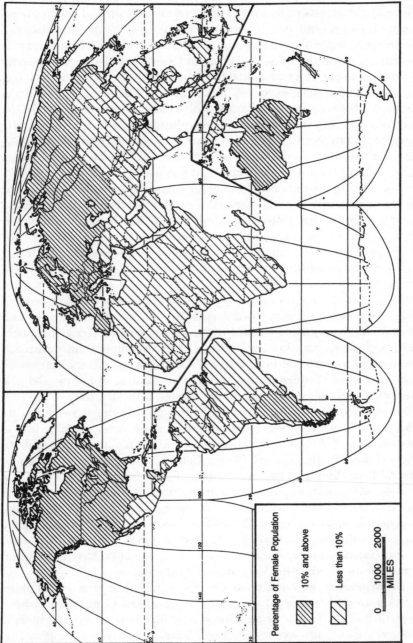

Figure 1.3 Global distribution of the female population aged 65 years and over

(*Network News* 1990: 9). These data remind us that older women represent an increasingly diversified group with different needs; those in their late fifties and sixties may still be in the labour force, whereas those in their late eighties and nineties are likely to be frail and to require considerable assistance to manage their daily lives. The chapters by Ruth Fincher, Sandra Rosenbloom and Patricia Sachs discuss in detail some of these differences in experience.

The increasing numbers of the oldest groups are especially marked. For example, while the population aged 55–64 grew by 20 per cent in Spain and Italy between 1950 and 1980, the number over eighty-five increased by more than 100 per cent. Most remarkable has been the expansion of the proportions of people living into their eighties and nineties. Since two-thirds to three-quarters of those over eighty-five in most western European countries are women, it is clearly important to learn more about the experiences and needs of very elderly women (*Network News* 1990: 9). Not only are high proportions of elderly women likely to be widowed, but more than 10 per cent of women currently aged fifty-five and over in many of the countries of the European Community (and nearly one-quarter in Ireland) have never married (*Network News* 1990: 10). How do these older women support themselves? What roles do they fulfil? How do families, the State and other institutions assist them? What is the nature of their geographies and how do these relate to their well-being? What are the differences within the region? A recent report on the *Social and Economic Situation of Older Women in Europe*[1] suggests some answers to these questions. For example, labour force participation by women aged 55–64 varies markedly among countries, from a low of 16 per cent in Belgium and the Netherlands to a high of 50 per cent in Denmark. Whether these employed women are likely to be married, divorced, widowed or never to have married also varies among the countries. Though older married women are generally less likely to be employed than unmarried women of the same age, in Belgium, Greece, the Netherlands and the United Kingdom it is the widowed and divorced women in the 50–64-year-old age group who are the least frequently employed. The kind of work that women do also varies among the countries. Agriculture is still an important source of employment for older women in southern Europe and Ireland, for example, though less so elsewhere. Economic changes are also modifying older women's employment prospects. Growth in the service sector and in part-time work seems to be associated with their increasing participation in Denmark, the Netherlands and Portugal, though the work does not pay well.

The ability of older women to manage their everyday lives clearly relates to both their incomes and their health, with the latter especially affecting their geographic mobility and capacity to fulfil their needs.

Surveys reported in the European study note that poor health inhibits women's ability to climb stairs, go shopping, use public transportation or walk to the nearest medical facility unaided. Sandra Rosenbloom addresses similar problems among older women in the United States in her chapter in this book. Though European Community and British Commonwealth nations generally provide national health insurance and health service schemes, differences remain in local provision, especially between rural and urban areas, and are heightened as emphasis shifts from institutional to community-based care. The consequences of this unevenness and some implications of elder-care policies for middle-aged women care-givers are addressed in this book by Ruth Fincher, who writes of the situation in an Australian metropolitan area.

Adequate housing for older women is another issue identified in the European study. Frequently their older homes lack adequate amenities, especially in such places as the rural areas of Italy and Greece. Some of the dimensions of the housing situation of older women in a rural community in the United States are dealt with by Patricia Sachs in her chapter which describes how women and their husbands have taken steps to improve their housing and to sustain a supportive neighbourhood.

The increasing numbers of older women in many Latin American and Caribbean countries in principle have access to various forms of state support, especially pensions. Changes in the global economy, however, threaten the abilities of debt-burdened countries to meet these obligations. Further, because coverage is generally limited to those in the paid labour force, the many women who are agricultural workers, domestic workers and informal sector traders are ineligible for pensions. Because coverage, age of eligibility, benefits and patterns of labour force participation differ among the countries of the region, the results for women also vary. For example, in the early 1980s, just under 25 per cent of El Salvadorean women were employed in positions with social insurance coverage, compared with almost 45 per cent of women in Barbados (Mesa-Lago 1990: 9). These data refer to women currently employed, however. The essays in this book by Janet Henshall Momsen, Lydia Pulsipher and Janet Townsend address the survival strategies of women who are now elderly in Latin America and the Caribbean; among other things they combine their own labour, remittances from adult children and help from other kin or neighbours, or move to live with adult children who have migrated away from their home regions.

Whether women have the power, status and autonomy to secure an acceptable quality of life in their later years, as opposed to simply attaining longevity, is a difficult question to answer. In her review of literature on a range of pre-industrial societies, Chaney (1990) suggests that women had a greater degree of power in the family and community in pastoral, nomadic and hunting and gathering societies than in agricultural and industrial ones. Still, she notes that prestige and influence also had to be attained (rather than simply being bestowed on the basis of age) in these societies, especially through the exercise of creativity and initiative, principally in service to the community. How older women in a former mining community exercise such initiative is the theme of Patricia Sachs's chapter. Lydia Pulsipher's account of the life histories of elderly Caribbean women and Janet Townsend's comments on older women in pioneer settlements in Colombia extend these explorations.

WOMEN'S GEOGRAPHIES FROM A LIFE COURSE PERSPECTIVE

In approaching the geographical experiences of women from a life course perspective, we are largely charting new territory. With few exceptions, feminist geographers have emphasized the behaviour, concerns and perceptions of women who are implicitly in the younger and middle years of adulthood, focusing on questions related to employment, the journey to work and childcare in their studies in Western societies and on the relationships between development and women's productive and reproductive work in Third World countries. They have paid almost no attention to the years of childhood and adolescence, and only a little more to the lives of older women.[2]

The larger multidisciplinary feminist literature provides more guidance in approaching our topic, though here again, studies which make the life course their central concern are relatively rare. Most notable is a collection of British essays (Allatt *et al.* 1987), which brings together discussion of conceptual issues with case studies that cover various life stages, dealing with both private lives and structural issues. As we noted earlier, the interest in the diversity of women's experiences in the feminist literature has emphasized differences based in race, ethnicity and class, as well as sexual orientation and cultural context. Only recently have we seen the call to consider all phases of the life course and development over the life span (Rosser 1991). Rosser locates the neglect in the composition of feminist scholars. Not only have they been predominantly white, middle class and Western, they have also been relatively uniform in age (20–45). She suggests that only as women's studies scholars began to age have we seen any serious attention paid to issues facing older women, such as menopause, osteoporosis and healthcare and housing for older women.

Despite its gaps, the feminist literature does include some studies that incorporate a life course perspective in their consideration of the ways in which adult women develop and change. A number of these focus on women's experiences of midlife as these reflect social changes in the

United States. The collection *Women in Midlife* (Baruch and Brooks Gunn 1984), for example, includes essays on the influences of economic conditions, ideologies and the normative timing of marriage and child-bearing on the problems women face, the quality of life they experience and the powers they assert. These essays also explore how other sources of difference (culture, race, cohort) interact with the phenomenon of midlife. Among other feminist works interpreting women's life course transitions are Mary Catherine Bateson's (1990) portraits of five 'success-ful' women shaping their professional and personal lives as they con-front losses and opportunities, and Myra Dinnerstein's (1992) study of married midlife women who have turned from homemaking to business and professional careers. She examines how they negotiate changing roles within the family as they develop new aspirations and self-images.

Returning to geography, this time to non-feminist literature, we find a few approaches of interest, though surprisingly little relevant empirical research. Population studies, largely in their examination of intra-urban residential mobility in North America, have long incorporated the concept of the family life cycle into their models.[3] This work has two major limitations from the perspective of understanding the lives of contemporary women, however. First, it has generally proffered a culturally and temporally bound notion of family structure, and second, it has portrayed women only in the context of family relationships. Much more attention needs to be paid to the dynamic nature of household units, as Clare Stapleton suggested over a decade ago (1980), and women's lives also need to be situated in a wider array of relationships and contexts.

Geographic research on environmental perception and behaviour has also been sensitive to life stage variations (Aitken *et al.* 1989; Saarinen, Sell and Husband 1982), though as Janice Monk (1984) pointed out, the stages are treated discretely, with few efforts to examine transitions or connections between the various phases of childhood, adulthood and old age. A rare exception is the comparison by Graham Rowles (1981) of his own work on the elderly with Roger Hart's work on children, in which he concludes that older people's perceptions of their shrinking behavioural space does not represent the inverse of children's learning about their expanding spatial realm.

We had hoped that the Swedish time geographical studies initiated by Torsten Hagerstrand and his colleagues would provide us with more fully developed insights to approach our undertaking. Conceptually, this research acknowledges longitudinal perspectives as well as people's short-term daily needs to coordinate movements in time and space. But while there are theoretical explications (for example, Pred 1981) our search yielded only a few empirical studies which go beyond the daily scale. Among the exceptions are Hagerstrand's study of weekly and seasonal movements (1982) and of lifetime moves (1975) in rural communities in Sweden, and Solveig Martensson's (1977) analysis of the complex interplay between the developmental needs of children and the shifting time commitments of their parents' work.

Having identified relatively little to guide us in these literatures, we turned to the multidisciplinary research on the life course, drawing particularly on writing in developmental psychology, family history and sociology of the family. Though the questions and approaches in these fields differ in a number of ways from our own, most notably by their lack of interest in the spaces and places of women's lives, we nevertheless find their conceptual insights valuable. First we should comment on basic terminology. The long-standing and relatively common term 'life cycle' is no longer widely used in this literature, especially in the United States. One reason for this retreat, Alice Rossi argues, is that 'cycle' inappropriately implies multiple turns (1980). A second, less obvious, reason is that the concept is associated with earlier research that often linked life experience to a relatively fixed or inevitable series of biological stages and ages (Allatt *et al.* 1987; Hareven and Adams 1982; Rossi 1980). Research that now emphasizes the diversity of experiences within an age group and the lack of clear associations between chronological age, perceptions and behaviour, has instead adopted the terms 'life course' and 'life span'. The former is commonly used by sociologists and historians, the latter by psychologists (Neugarten 1985; Rossi 1980). This research is primarily concerned with the transitions that occur as people age; it explores their pathways through the various structures in the major role domains of life. These transitions are interpreted within the context of changing historical conditions. Because our interests are closer to those of the historians and sociologists than to those of psychologists, who focus more on individuals and the inner world, we have chosen to use the term 'life course'. With the British sociologists Allatt *et al.* (1987), however, we find some continuing value in the 'ages and stages' idea, particularly as we recognize the importance of attending to demographic variations among women world-wide. Further, we are attracted by Bateson's recognition that women make many new beginnings in life, but in living new lives draw on their pasts even as they use them in new ways (1990). She invokes ecological concepts in proposing alternative ways of thinking about 'composing a life', among them the notion of recycling. As geographers, we place importance on the significance of space as well as time. Bringing together these ideas, we have titled our project 'full circles', choosing a metaphor that refers to space as well as time.

An important premise of the multidisciplinary literature is that change throughout the life course is based not only in biology, but in experiences of family, community and history. Since these experiences reflect multiple influences, which have interactive and cumulative effects, divergence among people increases over the life course (Campbell *et al.* 1985; Featherman 1983). It thus becomes important to examine the significance of prior experience on later life, both for individuals and for groups (Back 1980). In this vein Ozawa (1989) identifies many of the causes of poverty among contemporary women in the United States in their educational and employment histories and in the ideologies about women's economic support that have prevailed during their lifetimes. From a developmental perspective, these past experiences do not determine present and future courses, but they shape the capacity for choice and constrain options.

Within this framework it is important to identify the cohort to which a woman belongs. Behaviours we associate with a specific life stage may more truly reflect the conditions through which a group has lived collectively, such as its access to education, than biological age. As Rossi (1980) has pointed out, much of what has been written about middle age in the United States is based on the particular histories of people born in the 1920s and 1930s, whose childhood and young adulthoods were affected by economic depression and war, in ways not shared by those now in middle age. In an interesting study of Italian women Saraceno (1991) demonstrates clearly how three recent cohorts of women reveal very different patterns of labour force participation, varying in terms of their age of entry into the market and continuity of employment. These differences reflect changes in gender ideologies, marriage laws, educational provision, fertility trends and restructuring of the Italian economy. Changing conditions have not only influenced employment experiences but also family roles and the psychological burdens each cohort bears as the women in them try to balance reproductive demands and their aspirations with the employment scene. Jeanne Fagnani, Kathleen Christensen and Geraldine Pratt and Susan Hanson discuss these issues in their chapters.

The Italian study highlights one of the more interesting questions for scholars concerned with women and the life course – assessing how the conjunction in timing (or synchronization) of diverse roles affects experience (Hareven 1982). To understand the life course we need to be aware of both the issue of time, whether this is individual time, family time or historical time, and of the diversity of roles that women assume. Our examination of women's roles has to be extended so that we see them not only as mothers, wives and workers, but also in relation to other generations – as daughters, grandmothers, aunts and so on – and in domains outside the family and workplace, their wider community of friends as well as in relation to various social and political institutions (Lopata 1987), which will vary in significance over the life course. As geographers, we further argue that the study of temporal and social circumstances needs to be extended to consideration of the spaces and places within which women construct their lives.

THE GEOGRAPHY OF *FULL CIRCLES*

In his 1966 novel *Hopscotch*, Julio Cortazar developed an innovative structure: one could take the traditional route, reading from Chapter 1 forward, or one could play 'hopscotch' – starting in the middle and jumping from chapter to chapter, following his notation at the end of each chapter. In organizing *Full Circles*, it was tempting to do the same. The intersections between different chapters are rich and provocative, and each strategy for structuring the book seemed to close as many connections as it opened. In the end, we chose to move more thematically than chronologically or geographically, though not surprisingly, both time and space play a role in the organization of the book. Hopping across the structure we have adopted may illuminate other themes as well.

Although part of our intent in examining women's experiences across the life course was to get away from the predominant practice of situating women at the nexus between production and reproduction, this positioning has proven obdurate in defining women's lives across history and geography, and occupies many years over the life course. Moreover, the State and other institutions generally have proven tight-fisted in supporting structures, policies and practices that might lighten women's burdens. Patriarchy and global capitalism, not coincidentally, are both behind and buttressed by the persistence of these relationships that exploit women's labour in the two realms. Many of the strategies for confronting these enduring forms are local and vary widely from place to place, as the chapters of this book attest, but analytically they represent women's engagements with similar social relations and political-economic structures. Though the strategies vary geographically, historically, by class, cohort, nationality or ethnic group or at the personal level, the different chapters reveal ways in which they are part of a single process. The chapters are intended to work both with and against each other to cross-fertilize and inform theory, practice and policy at all geographic scales. To connect these practices in ways that intensify their effect will require more than a game of 'hopscotch'. Nevertheless, we invite you to 'play', as an initial approach, to make the most fruitful connections between the chapters.

We begin with two chapters that explore some of the spatial problems associated with women's positioning at the cusp of production and reproduction in the United States. Geraldine Pratt and Susan Hanson prise apart the notion of working woman by examining the relationships

21

between domestic responsibility, marital status, employment characteristics and time in the life course, along with the role of space and place in the mediation of the often conflicting demands of home and work. While Pratt and Hanson suggest that the spaces of married women are the most constrained because of their heavy family commitments and therefore that, among other strategies, they seek employment closer to home or have a more contingent relationship to work than do men or other women, Kathleen Christensen's chapter is focused on home-based work as a resolution to this situation. While many of the women she studied determine that home-based work is the best solution to the competing demands on their time, the life course implications of Pratt's and Hanson's findings are witnessed among middle-aged and older women participating in Christensen's research. That is, while many of the women who choose to establish themselves as paid workers or entrepreneurs in their homes are no longer bound by family responsibilities, it is frequently the case that they feel they have little choice other than such work because after their years at home rearing children, they are either insufficiently appreciated (or actively denigrated) by employers, or feel uncomfortable negotiating commuting or the workplace itself. They therefore decide to engage in home-based work. The cohort effects on contemporary older and middle-aged women in the United States are revealed in both chapters in narratives detailing women's socialization to be wives and mothers first, as are their difficulties in or tangential relationship to the workplace because of this. These patterns of socialization die hard, as one of the younger women speaking in Christensen's chapter reveals, and many younger women still experience the tug of home-based responsibilities to a greater extent than their male partners. Yet one of the most important lessons of Christensen's research, and of the life course perspective in general, is that women construct and can change their lives, even against the insistent structural constraints associated with capitalist and patriarchal social relations. The narratives in Christensen's chapter illustrate some of the strategies by which women alter their lives and rearrange or dismantle patterns set at earlier phases of the life course.

The question of socialization leads to the next chapter, by Cindi Katz, which examines the relationship between girls' socialization and spatial range in two discrete sociocultural settings, both to draw comparisons between the two and to scrutinize the lasting effects of girls' limited spatial experience compared to boys'. Katz's chapter points out how in both the United States and Sudan the limits set on girls' movements and their internalization of fear for their safety results in a diminution of their autonomous movement in many of the sociophysical settings of their lives. She observes that the negative repercussions of this deprivation over the life course accrue to the women themselves, as well as to the society as a whole. Much of Katz's work is focused on rural Sudan where she notes that, somewhat counter-intuitively, girls have greater freedom to roam their environment than they do in most places in the United States.

Katz's discussion of children's relative autonomy within the houseyard complements the work of Lydia Pulsipher and Janet Momsen who discuss the sociospatial dynamics of the Caribbean houseyard and their relatively liberating aspects for women. Pulsipher's chapter interprets the experiences of older women to reveal the considerable autonomy they enjoy over the years in their multifaceted intergenerational social relationships with other women as well as men in lives centred on the houseyard. She demonstrates that the communal nature of the yard enables women to bear and raise children outside of marriage, and to participate in productive activities, whether nearby or in distant locations. This flexibility is facilitated by the houseyard structure, within which relatives, both male and female, may take care of one another's dependants for short periods or years at a time. The give and take between those who remain in the houseyard and those who emigrate and send remittances is also explored by Henshall Momsen, who addresses the multiplicity of economic roles held by women in the poor, semi-proletarianized small farming population of the Eastern Caribbean islands of Nevis, Montserrat and Barbados. She looks at the variations in the nature of work and in the labour force participation on each island for women at different stages of the life course, and addresses some of the contrasts between men and women, relating these to the level of economic development, occupational structure, the resources available to each group and their various domestic responsibilities. In discussing the differences between the agricultural, industrial and service sectors as sites for women's labour force participation, Henshall Momsen notes some of the life course implications of certain policies such as the provision of education or the availability of pensions. She demonstrates how access to such resources will affect the way women at different stages in the life course participate in the larger economy, and the significance of these articulations for the future.

The next chapter focuses on some of these issues with particular reference to communities along the edge of the forest that demarcates the frontier of settlement in Colombia. Janet Townsend addresses the profound changes in women's everyday lives when they migrate to these communities from other rural areas. These include erosion in women's rights, the narrowing of their work experience and their often extreme isolation. She traces the paths by which women withdraw or are excluded from the productive labour force and either become housewives, and thus dependent on others, or migrate elsewhere, either in search of work or to accompany children in school or in the labour force. Townsend's chapter traces the life course of the pioneer settlements as well, pointing out that those who move to them after the settlements have been established for some time will have more constricted opportunities than those who come in the earlier years.

The evolving life course of a community and the intersections between men's and women's work are the themes of Patricia Sachs's chapter which follows. Sachs, studying an Appalachian coal mining community following closure of the mine, addresses how people can remain in place and redefine their lives through new forms of labour and redefinition of home and community. The process she discusses is almost the antithesis of the establishment of the pioneer community; whereas in the frontier community Townsend documents sharpening distinctions between men's and women's roles, in this community suffering the loss of its productive base, Sachs describes a blurring of the gender division of labour. She discusses how those who had known no other home than the company town found it difficult to leave all that was familiar, to live and work elsewhere. Thus older residents stayed after the mine closed, occupying themselves with the improvement of their homes and neighbourhood and living modestly on earnings from sporadic work, pensions, public assistance, black lung benefits and the produce of their gardens. For those affected by the mine's closing at an earlier stage in their life course, there was little alternative than to migrate elsewhere in search of work or new opportunities.

The question of residential decision-making is at the heart of the next chapter by Jeanne Fagnani. Working in a very different setting, among upper-middle-class families in Paris and its suburbs, Fagnani discusses the concerns that dual career couples at midlife weigh as they contemplate a residential move. Her chapter explores not only the competing demands of two careers and various household requirements, but also looks at the ways in which sociocultural values shape residential choice. Echoing the issues raised in the first chapters, Fagnani finds that women with children are more tightly bound to their domestic responsibilities than men, and orient many of their residential decisions around their children's well-being and their attempts to juggle multiple roles at home and in the workplace. In pointing out some of the conflicts in these women's lives, Fagnani recognizes their much greater class capacity than most to negotiate these, thereby suggesting the nature of the difficulties for women with fewer options.

The question of women's multiple roles is central to the chapter by Damaris Rose which examines state policies concerning childcare and their articulation with class, ethnicity, family structure and life course position in three neighbourhoods in Montréal, Canada. Rose looks at the ways in which women negotiate the available childcare options according to these other circumstances. She addresses the various time–space constraints of parents with young children, especially those with children in daycare and school, demonstrating how the production of urban space reinforces uneven socio-economic patterns and the gendered division of labour. Her chapter also reminds us that the constraints that have to be negotiated to meet children's needs change as the children grow up.

These socio-economic patterns and prevailing ideologies produce particular forms of mobility and access to services for women and others that vary throughout the life course. These themes are addressed directly in the next chapter by Sandra Rosenbloom, who looks at the nexus between employment, household responsibilities and transportation choices at various life stages for married and unmarried women in the United States and Europe. As with several of the authors, Rosenbloom finds employed mothers with employed partners more likely than either their partners or single mothers to combine their responsibilities in the spheres of production and reproduction, through such actions as linked or multiple journeys. She also addresses, as does Rose, the differences in mothers' experiences and needs according to their child's age. Additionally, her comparison of the situation in the Netherlands and the United States illuminates the significance of such contextual effects as the prevalence of different modes of transportation.

Rosenbloom's chapter also uses a life course perspective to examine the travel patterns of elderly women in the United States. She discovers that the majority of the women in the cohort presently over seventy years of age lack driver's licences and that their mobility is often severely curtailed after the loss of a partner, though she points out that it is the failure to hold a licence, rather than marital status, which is critical. She offers the prospect of older women's mobility patterns changing in the future because more recent cohorts include much higher proportions of women who hold licences. These women are likely to be more mobile than their predecessors, until they are faced with declining abilities.

The isolation of older women is one of the problems at the root of the relationships explored in Ruth Fincher's chapter. She examines how particular state policies in Australia implicitly privilege certain of women's roles – wife and mother of young, able-bodied children – and thereby may exacerbate difficulties women encounter in coping at other points in the life course or when they are outside these roles. In concert with the class and ethnic nature of Australian society, these policies, which particularly position women as mothers of young children, are at once the outcome and reinforcement of socio-economic and gendered imbalances in the relations of production and reproduction which tend to obscure the fact that women in their 'middle years' are labour force participants. Fincher explores how general changes in the life course in

Australia, such as the postponement of child-bearing and the ageing of the population can create new and contradictory demands on women in their middle years who are caring simultaneously for children and ageing relatives. She illuminates how women of different class and ethnic backgrounds vary in their capacity to respond to these imbalances and to the policies that underlie them. Fincher is careful to point out that the State is not in itself producing these social constellations but is in part responding to women's activism and political power. To some degree, the policies reflect the interests, life course stage and experiences, class and socio-economic concerns of women in government and high-level decision making positions.

With Fincher's careful attention to how the State writes women's lives, we have indeed come full circle to the very question raised by Geraldine Pratt and Susan Hanson. Each of the chapters in this book attempts to pull apart some of the ways in which women's lives are constructed in theory and practice within the dominant relations of production and reproduction. Recognizing both the multiple determinations in women's lives and women's historical agency, it should be impossible to 'script' them at only one point – frozen – as wives and mothers, and to have all policy and practice revolve around this positioning. It has been our intent to contribute to these deconstructions and expose differences of age, class, ethnicity, nationality and individual values, as well as those of geography and social history, to open paths towards change that begin in all of the many 'locations' in which women find themselves and which go 'full circle' to address the webs of sociocultural and political-economic relations in their lives.

2

WOMEN AND WORK ACROSS THE LIFE COURSE

Moving beyond essentialism

Geraldine Pratt and Susan Hanson

We women have lived too much with closure: 'If he notices me, if I marry him, if I get into college, if I get this work accepted, if I get that job', there always seems to loom the possibility of something being over, settled, sweeping clear the way for contentment. This is the delusion of a passive life. When hope for closure is abandoned, when there is an end to fantasy, adventure for women will begin.

Heilbrun (1988: 130)

Observing the lives of several well-known women writers, Carolyn Heilbrun notes that they are patterned differently from those of men. Many of these women created themselves as writers in their middle ages: 'We must recognize what the past suggests: women are well beyond youth when they begin, often unconsciously, to create another story' (Heilbrun 1988: 109).

When one examines the narratives constructed, within and outside of the academic world, about women and their lives of work, one gets little sense of these multiple stories. Waged employment taken up by women is often portrayed in uncomplicated terms, as uniformly insecure, poorly paid and without benefits and career mobility.[1] Women's domestic responsibilities figure prominently in explanations of women's subordinate labour market position. But these explanations rest on an overgeneralization of one life path (one that involves a woman marrying and then bearing and caring for children) and one period within that life course (the period of intensive care for young children). Human capital theorists, for example, search for the rationality of women's low wages and occupational selection in their movement in and out of the labour force during child-bearing years (for a recent critique of this application of human capital theory, see Pratt and Hanson 1991a). Some male trade unionists have legitimized women's low wages and their own claims to a 'family wage' through their selective portrayal of women as mothers of small children and dependent wives (who, if in paid employment, work

27

only for 'pin money') (Cunnison 1987). Anticipating that women will leave employment for child-bearing/rearing, employers have indiscriminately slotted them into certain low-wage, dead-end jobs (Peterson 1989); similarly employers have justified the exclusion of all women from certain job categories on the grounds that the job poses a danger for foetuses. For example, in March 1991, the US Supreme Court ruled in *United Auto Workers* v. *Johnson Controls* that employers cannot bar women from certain jobs on the grounds that such jobs involve exposure to substances dangerous to potential offspring. Though more critical of the outcome, some feminist theorists, for example Beechey (1977), also build their explanations for women's low wages from the stereotype of the married, dependent woman (for a critique, see Redclift, 1985). 'The family', and particularly the young family, has played a leading role in feminist explanations for women's marginal position (in terms of hours worked, continuity of employment and status and conditions of work) in the labour force (Barrett and McIntosh 1982).

'Any cultural description', writes Clifford, 'is an ensemble of anecdotes, narratives, interpretations, typical characters, allegories, partial arguments, in short, a complex rhetorical performance' (quoted in Elshtain 1987: 320). While marriage and children have a non-trivial impact on many women's (and men's) lives, one that reverberates throughout them, a description that highlights a single period in one type of life course is clearly partial. This description has played an important role within feminist critical (as well as human capital theorists', male trade unionists' and employers') rhetorical performances, but its partiality now seems more obvious and increasingly problematic. This is so for a number of empirical, theoretical and political (strategic) reasons.

First, child-bearing absorbs fewer and fewer years of the average woman's life and there is now considerably more flexibility in the timing of these years than in the past (Davis 1988: 72; Phillips 1987: 60; Stacey 1990: 9). Second, 'the nuclear family' and women's dependency on a male wage-earner seem to describe the experiences of fewer and fewer people. In 1950, three-fifths of households in the United States took the form of a breadwinner and full-time homemaker; by 1985, 7 per cent fitted the pattern of breadwinning father, homemaking mother and one to four children under the age of eighteen (Stacey 1990: 10, 285).[2] The time that women spend out of the labour force to care for their children has shrunk: by 1988, 54 per cent of US mothers with children under the age of three had paid employment, an increase of 32 per cent from just eight years before (Stacey 1990: 15). The idea that women are working for a secondary wage is increasingly untenable, if only because of the increasing instability and decline in wages of traditionally male blue-collar jobs (McLaughlin *et al.* 1988; Stacey 1990). Family composition is extremely varied. By 1988, one out of every four US children

28

lived with a single parent (Stacey 1990: 15). This statistic only begins to signal the fluidity and complexity of US families, which also take blended, intergenerational and communal forms. Stacey coined the term 'postmodern family' to convey the sense of variable family strategies: 'An ideological concept that imposes mythical homogeneity on the diverse means by which people organize their intimate relationships, "the family" distorts and devalues this rich variety of kinship stories' (Stacey 1990: 269).

There are also theoretical and strategic reasons why feminist scholars should choose not to impose this homogeneity on kinship arrangements and the multiple stages in an individual woman's life. One flows from a careful reading of how employers and male employees have attempted to exclude women from particular jobs by over-generalizing their difference, as defined by child-bearing capabilities. Reskin (1991), for example, argues that men's higher wages rest, at bottom, not on occupational segregation, but on men's desire to retain their privileged position, which in turn rests on their defining women as different; men seek, she maintains, to preserve male–female difference, and they do this by pointing in particular to 'women's "natural" roles as wife, mother' (p. 150). At the core of policy debates, too, is the question of whether public policy should treat male and female workers as the same, or recognize differences, with the chief source of difference mentioned being women's greater domestic workload and their child-bearing capability.[3]

Posing the question in this way not only assumes an unrealistic stability in domestic demands across the life course, but also treats all women as if they were caught indefinitely in the child-bearing years.

While feminists are unlikely to reproduce this particular biological essentialism (that of collapsing the category 'woman' with a temporally constrained set of biological conditions that may not even be relevant to every woman of child-bearing age (e.g., those who choose not to bear children), see Smith 1983), we have tended to over-generalize the links between sex and a specific set of household relationships in our explanations of occupational segregation. This has led to a possibly essentialist link between gender and a particular set of occupations (Siltanen 1986). Siltanen has attempted to redirect feminist enquiry into inequality in the occupational structure by demonstrating that sex *per se* is less important as a condition for entry into a 'female-dominated occupation' than a particular set of domestic responsibilities (which may be held by men as well, though typically are not). In other words, she is suggesting the need to specify more closely the particular domestic experiences that lead one to take employment in what she calls 'component wage' (as opposed to 'full wage') jobs, and to disentangle this theoretically from the categories woman and man. This not only sharpens an explanation of occupational segregation, it provides a framework for understanding differences among women, as well as common ground between some women and men.

The need to deconstruct the category woman is now widely acknowledged within feminist theory. Much of the discussion about difference has moved around the experiences of class, race, ethnicity, religion and sexual alliance, with very little consideration given to the ways in which women's experiences change through the life course.

Attention to change through the life course is important because it allows us to contain and contextualize debates about equality and difference. It allows us more carefully to articulate the processes that link many women to the secondary labour market. By reading change in context, we can sharpen our critique of existing linkages between domestic and paid work that characterize one phase of some women's life course, and explore more carefully how this phase resonates through the life course. At the same time, a life course approach suggests a more careful analysis that admits both the diversity of women's experiences and social change. Most importantly, this framework reminds us that gender is socially constructed, that the category woman is created and recreated in different ways throughout even a single woman's life. Individuals may change, in their personal values and aspirations, in the constraints placed upon them or opportunities available to them; as Carolyn Heilbrun notes, in their middle ages some women create another 'story'.

Lives are lived through time; they are also lived in place and through space. This has been recognized by those who study women's lives through the life course when they notice, for example, that the local labour market in which an individual lives has a tremendous influence on opportunities for paid employment (Bird and West 1987; Tilley 1985; Treiman 1985). As a generalization, however, the spatiality of social life, as it intersects with temporal change, has not been well explored. This seems a significant omission, because it is likely that typical life course paths vary considerably from place to place, given differences in employment structure and cultural values concerning domestic work and ageing.

Geographers' interpretations have tended toward the opposite bias by seeing women's work through the lens of place and space, with insufficient sensitivity to temporal change. In this chapter, we attempt to hold the two interpretative frameworks in tension, in order to explore more fully women's lives of work in Worcester, Massachusetts. In particular, we attempt to break into the category 'woman' to demonstrate how position in the occupational structure is conditioned by domestic responsibilities, marital status and time in the course, and how space and place are critical to the mediation of home and work. Taking

30

place seriously, we begin with a brief description of the Worcester context, against which our remarks about women and work can be situated.

WORCESTER, MASSACHUSETTS: CONTEXT AND DATA SOURCE

Worcester is an old industrial centre, well known for the production of industrial abrasives, wire, textiles, leather products, envelopes and jet engines. Since the late 1960s, manufacturing has been renewed in Worcester, through the growth of high-technology firms that manufacture goods such as semi-conductors and optical products. In 1980 manufacturing still accounted for almost one-third of all jobs in metropolitan Worcester, as compared to an average of 22 per cent across all metropolitan areas in the United States. Still, Worcester's economy is restructuring: by 1988, manufacturing accounted for only 22.5 per cent of jobs in the Worcester region, the proportion of jobs in the manufacturing sector had declined from 46 per cent since 1970, and economic growth was in the service sector, especially in the areas of insurance, healthcare, higher education and retailing.

Given the growth in service industries, it is not surprising that women participate in the labour force at high rates. Sixty per cent of women in New England were employed in the labour force in 1987 – a jump from 55.7 per cent in 1982 – a rate which was slightly higher than the national average of 56.0 per cent in 1987 (French 1988: 22). At the time that we conducted our household survey in Worcester, the unemployment rates were very low. In Massachusetts, in 1987 the unemployment rate was lower for women than men, at 3 per cent and 3.4 per cent, respectively (French 1988: 22), and there was an extreme shortage of entry-level consumer services, health and clerical workers.

Industries and occupations in Worcester are highly segregated by sex, as shown in Tables 2.1 and 2.2. As Table 2.2 shows, more than half of all employed women hold jobs in female-dominated occupations and almost two-thirds of the male work force are in male-dominated ones. Female- and male-dominated occupations have different reward structures: jobs in female-dominated occupations yield the lowest hourly wages and those in male-dominated ones the highest, a difference that remains statistically significant when the earnings of only women in these groups are compared (Hanson and Pratt 1991).

Low wages pose a real problem for households trying to navigate Worcester's housing market. From 1977 to 1986, house prices in Worcester tripled, and in the second quarter of 1986 Worcester was ranked as the fifteenth most rapidly appreciating housing market in the United States (Pratt and Hanson 1991b). (However, the inflation of

Table 2.1 Gender differences in industries of employment

Industry	Percentage of workers		Percentage of gender labour force	
	Women	Men	Women	Men
Distributive services (e.g. transportation, communications, public utilities, wholesale)	41.0	59.0	4.8	12.0
Health, education, welfare (e.g. medical services, educational institutions, public administration)	77.4	22.6	40.8	20.8
Consumer services (e.g. retail stores, personal services)	78.9	21.1	22.3	10.4
Producer services (e.g. banking, real estate, consulting services)	67.1	32.9	17.0	14.6
Construction	19.0	81.0	1.2	8.9
Manufacturing (e.g. textile products, metal industries)	42.3	57.7	14.0	33.3

Source: Personal interviews, Worcester, Massachusetts, 1987.

Table 2.2 The gender composition of the occupation types

	Occupation type[a]			
	Female-dominated	Gender-integrated	Male-dominated	Row totals
Women	181 (53.9)[b]	126 (37.5)	29 (8.6)	336
Men	12 (6.3)	55 (28.9)	123 (64.7)	190
Column Totals	193	181 (36.7)	152 (34.4)	526 (28.9)

a Female-dominated occupations are those three-digit 1980 census occupation codes in which at least 70 per cent of the incumbents, nationally, were women. Male-dominated occupations are those in which at least 70 per cent of the incumbents were men. All other occupations are defined as gender-integrated.
b Figures in parentheses are row percentages.
Source: Personal interviews, Worcester, Massachusetts, MSA, 1987.

house prices has stopped and even reversed since the downturn in the economy, starting in 1989.) Families compensated for high house prices to some extent by sharing housing across generations (Pratt and Hanson 1991b). This may account for (or perhaps reflect) the remarkable stability of Worcester households (Hanson and Pratt 1988).

This stability may also reflect the history of Worcester as a city of ethnic communities. In the early-to-mid-twentieth century, the Swedish, Irish, Greek, Italian, French Canadian and Yankee communities, among others, flourished (for accounts, see Gage 1989; Rosenzweig 1983; Warshaw 1991). Since 1980, there has been a large influx of Latin-American and Asian immigrants (Warshaw 1991).

In our discussion of women's work experiences, we draw upon two survey data sets. In 1987 we interviewed women and men in a representative sample of 620 Worcester-area households. The interviews consisted of both structured and open-ended questions, aimed at finding out how people decide where to live and where to work, how they come to do certain types of work, how they find their jobs and how they evaluate specific aspects of their work and home environments.[4] In 1989 we interviewed employers and employees from 143 manufacturing and producer services firms in the Worcester area. This sample was structured by area. We narrowed our geographical focus to four areas within the Worcester region: two city and two suburban ones. (For a description of these areas, see Hanson and Pratt 1990.) One of our objectives was to examine the extent to which employers try to tap distinctive, spatially-constrained pools of labour. We obtained a considerable amount of information concerning employers' preferences *vis-à-vis* employees and their labour recruitment strategies.

MARRIAGE, DIVORCE AND PAID EMPLOYMENT IN WORCESTER, MASSACHUSETTS

In her study of Australian secretaries, Rosemary Pringle (1989) claims that successful 'executive' secretaries, as compared to other secretaries, have very distinctive family circumstances: they are single, separated, divorced or widowed. Two surveys conducted by American executive search firms highlight the different life course trajectories lived by male and female corporate officers. Among the men surveyed, 99 per cent had married, 95 per cent were fathers and only 4 per cent were separated or divorced. In comparison, 20 per cent of the women had never married, another 20 per cent were divorced or separated and more than half were childless (Fuchs 1989).

We can tell a similar story about women living in Worcester, and in our case we tell it about women in all occupations.[5] We looked first at women earning relatively high incomes, over $35,000 a year. There were very few of these women to investigate: of the 335 employed women in our sample, only eleven (or roughly 3 per cent) had incomes over $35,000. To put this into perspective, an income of $35,000 or more is not rare among men in Worcester – over 34 per cent of men with whom we spoke were in this income bracket. When one looks at the family circumstances of women earning high incomes, a very interesting pattern appears (Table 2.3). Almost all have what can be termed 'non-traditional' family circumstances: they are single, divorced or married without children. Only one woman in our entire sample was juggling child-rearing with a relatively well-paying job.

Women in male-dominated occupations are also unlikely to conform

Table 2.3 Family circumstances of employed women

	Women with income ≥ $35,000 per annum		Women in male-dominated occupations		Family circumstances of all employed women	
	No.	(%)	No.	(%)	No.	(%)
Single	1	(9)	3	(10)	21	(6)
Currently divorced, separated	4	(36)	6	(21)	32	(10)
Married, no children	4	(36)	2	(7)	78	(23)
Widowed	0	(0)	4	(14)	13	(4)
Husband ill, disabled	0	(0)	2	(7)	n.a.	(–)
	9	(81)	17	(59)	144	(43)
Married with children						
* Helping husband	0	(0)	4	(14)	n.a.	(–)
* Professional status attained before marriage and children, career on hold	0	(0)	3	(10)	n.a.	(–)
* Children grown	1	(9)	1	(3)	n.a.	(–)
* Children school age or younger	1	(9)	4	(14)	n.a.	(–)
	2	(18)	12	(41)	191	(57)

Source: Personal interviews, Worcester, Massachusetts, MSA, 1987.

to the 'norm' of marriage and children. Furthermore, the majority of women in male-dominated jobs who are married and have school-aged or younger children also have what might be called a 'fragile' relationship to paid employment. A number of these women saw their employment primarily in terms of helping their husbands. One woman in this group, for example, was a wallpaper hanger who helped her husband in his wallpapering and painting business. This is how she described her labour market participation:

'I can anticipate what he needs in a way that other people might not do. I take pride in his work, and I want to make it easier for him. So basically [I do this because] it's the time I can work and I enjoy working with him. I give him speed.'

To this woman, what was important about her job was that she was 'helping him . . . serving him'. Several other married women in male-dominated occupations with children of school age or younger mentioned that their careers were more or less 'on hold' at the moment, because they were presently working outside of the home on a part-time basis. In the words of an architect: 'It's been my choice to have kids as a priority. My profession has suffered because of it. My classmates are already far along and I'm still working part-time.' What we have found, therefore, is that women who are exceptional in their paid employment also tend to have 'exceptional' family circumstances.

A close examination of the career trajectories of divorced women reinforces this point. As Elizabeth Bird and Jackie West (1985) note, there is no single effect resulting from a family crisis. Ten of the forty-one divorced women in our sample were not employed (this compares to 30 per cent of married women) and eight of the ten received AFDC (Aid for Dependent Children) benefits. Half of AFDC recipients were using the benefits strategically, however, as a way of retraining to improve their employment opportunities. In the words of a thirty-seven-year-old woman with three children, aged eight, ten and seventeen, who had last worked as a waitress between 1976 and 1978 and was divorced three years before our interview:

'I liked waitressing. It was good quick money and I didn't have to claim tips. I was supporting my family even though I was still married. My ex- was working but I didn't see the money. [He was an alcoholic.] I ended up leaving him. We had a horrendous home life. He had a breakdown – both emotional and physical. . . . I was scared. What was I going to do with the rest of my life? How would I support my kids? I was tired of waitressing. I decided to go on AFDC, go to school part-time to get a degree for a longer-term career. I could be home for the kids in the mean time. It was a survival decision. I had worked in a cerebral palsy clinic as a teenager. I worked with handicapped children in school. So I looked into studying occupational therapy [which she was now doing]. I have been concentrating on a plan – a goal – to be a mother but go part-time to school to work towards the future.'

Of the thirty-one employed divorced women in our sample, eight mentioned that their divorce had forced them to return to the labour force and another eleven mentioned changes in their jobs: to improve benefits, for greater job security, from part-time to full-time hours, or to accommodate residential moves associated with the divorce.

Women who head households are less likely to work in female-dominated jobs than are married women[6] (Table 2.4), reinforcing Siltanen's (1986) point that many of the occupations labelled as female-dominated are better conceived of as 'component wage' jobs, ones that comprise just a component of a household's income. The majority of women who head households on their own do not have these occupations. Further substantiating Siltanen's argument, the personal incomes of employed divorced, separated and widowed women also tend to be higher than those of married women. For example, 48 per cent of married women had personal annual incomes below $9,000 while 'only' 33 per cent of divorced, separated and widowed women had incomes in this range. Alternatively, almost one in ten divorced, widowed and separated women had personal annual incomes of at least $35,000, compared to only one in fifty married women.[7] Income differences also reflect the fact that employed divorced, separated and widowed women are much more likely to work full-time in paid employment: this was the case for 82 per cent of women in these circumstances, as compared to 59 per cent of employed married women.

Table 2.4 Percentage of employed women in each occupational type, by marital status

		Occupation Type		
	N	Female-dominated	Gender-integrated	Male-dominated
Married or living with someone	270	57.1	36.3	6.6
Divorced, separated, widowed	45	40.0	40.0	20.0
Never married	21	42.9	47.6	9.5
$X^2=11.34$, d.f.$=4$, p$=0.02$.				

Source: Personal interviews, Worcester, Massachusetts, 1987.

Women who head households also want different things from their paid employment than do most married women. They care more about opportunities for advancement and good benefit packages.[8] Roughly similar proportions (about one-third) of employed married and divorced women (in this latter category we include separated women and widows) expressed frustrations about getting a preferred job, but the divorced women were much more likely to feel that they had hit the 'glass ceiling' because they lacked the requisite education; almost half (44 per cent) felt this way (compared to only 22 per cent of married women). Perhaps because of this, over one-third of divorced (and separated and widowed) women were currently upgrading their job qualifications by attending school, the majority on a part-time basis; this far surpasses the proportion of married women (22 per cent) currently enrolled in school. From a woman studying to become a registered nurse: 'If I'd stayed married I would probably still be waitressing. Now I am raising my son on my own. It makes it necessary to have a higher income and stability.'

Divorced women are, of necessity, more attached to the labour force. They value different job attributes, and many upgrade their qualifications to meet their occupational objectives; they are much less likely to have 'traditionally' female occupations. Over-generalizations drawn from married women's lives seriously distort divorced women's experiences, while simultaneously obscuring the linkages between 'traditional' family arrangements and occupational segregation. Five married women told us that they were frustrated from obtaining their preferred job because they 'were not allowed to'. No divorced woman told us that!

EMPLOYERS' THOUGHTS ON FAMILY AND WORK

Employers also play a role in shaping gender-based occupational segregation, by deciding that some jobs are 'naturally' suited to women and others to men. Strongly-held convictions about the influence of particular family circumstances on employee productivity could further segment employment opportunities among women.

A number of employers of manufacturing and producer firms that we interviewed in Worcester[9] did seem to know more about the family circumstances of their female employees than of their male workers. We asked them for information about the marital status of the women and men working for them. Most interviewees relied upon their memories rather than referring to firm records; for 11 per cent of the firms, the interviewee could provide this information only for women employees. This may reflect a number of factors: the very marked degree of gender-based job segregation in most firms interviewed and the fact that many women work as clerical staff, with closer personal interaction with management; the greater likelihood that women with children will require some accommodation from the employer to care for sick children; and perhaps a greater propensity for women to bring this aspect of their 'private' lives to public attention at work (Pringle (1989) remarks upon this fluidity between 'private' and 'public' among 'working-class' secretaries).[10]

It is interesting, however, to pair this seemingly greater knowledge of women's marital status with the fact that employers are more outspoken about their preferences regarding the marital status of female, as compared to male, employees. Roughly twice as many evaluative comments were made by employers about women's, as compared to men's, marital status, and their preferences *vis-à-vis* marital status vary by gender (Table 2.5). Employers who were prepared to make evaluative statements about men's marital status expressed a preference for married men, whom they viewed as more reliable and responsible. Only one employer preferred single male workers, in this case because single men are more willing to do overtime work. Preferences about women's marital status were more variable, depending on the occupations under consideration. Almost half of those employers who expressed an opinion preferred married women because of their stability. A slight majority, however, preferred single women or single parents. Their reasons for this were varied, but several pointed to greater acceptance of single women in gender-atypical or gender-integrated occupations.

Despite their stated preference for single women for some categories of jobs, it seems that employers are very sensitive to the location and spatial constraints of married women. This may in fact reinforce the linkages between a specific set of domestic responsibilities and occupational segregation. Kristin Nelson (1986) argues that employers of firms

37

Table 2.5 Employers' opinions about the marital status of female and male employees

Women employees	N
Preference for single women	10
Reasons given:	
• hours too erratic for married women	
• less likely to get pregnant and leave	
• job requires travelling	
• those with families to support could not afford low wages paid	
• single women better	
– in sales	
– in highest-ranking positions	
– in professional positions	
Preference for single parent	1
• 'type A personalities'	
Prefer **not** to hire divorced women	2
• more absenteeism	
• too busy with social lives	
Preference for married women	8
• mothers are good workers	
• married women are more stable	
	21

Male employees	N
Preference for married men	10
• more reliable	
• single men irresponsible	
• company likes married men with debts	
Preference for single men	1
• will do lots of overtime	
	11

Source: Personal interviews, Worcester, Massachusetts, MSA, 1989.

with large proportions of clerical occupations have identified suburban white, middle-class, married women with children as an especially attractive labour force. These women tend to be well-educated but relatively undemanding with respect to wages and job benefits. Given familial responsibilities, this prospective labour force is unwilling to travel great distances to paid employment. Nelson (1986) presents some evidence that suggests that employers in the San Francisco area are locating back-office operations in suburban areas in order to take advantage of this labour force of married women with young children.

We have also found that many employers are very sensitive to the availability of inexpensive and unskilled labour when they make their decision to locate their firm, and they tend to see this labour in gendered, familial (and in some cases, racial) terms. One-third of the employers interviewed stated that accessibility to an inexpensive labour pool was an important or very important factor dictating the decision to move to their present location, and almost half of the employers mentioned unskilled labour as being important to the firm's locational decision. Many employers seemed aware of the spatial immobility of this type of worker, as well as the family responsibilities that lie behind it. For example, a small manufacturer of wire cable told us that 'access to cheap labour was crucial to our decision to locate here ten years ago. We rely on unskilled, marginal workers, women with kids, high school students, and retired people.' All eight of his production workers are women, most likely married ones, because they all worked part-time and received no job-related benefits. The part-time inexpensive worker was also the target of this owner of a cleaning service: 'We are looking for cheap labour, unskilled labour. All our labour force are women and all work part-time.' From a third employer: 'The people we employ won't go more than five miles. There are only two males on the premises [sixteen women are employed]. Most are married women working for second incomes; a lot work during school hours. Fifteen miles is like global exploration to these people.'

This evidence is just suggestive, but it does indicate a symbiotic spatial fit between the needs of married women who have heavy domestic responsibilities that dictate that they must find paid employment close to home and employers who have identified this group as an ideal labour force. Though the employed married and divorced women whom we interviewed spent the same amount of time travelling to their current jobs (about fifteen minutes), divorced (and separated and widowed) women expressed a willingness to travel longer to a hypothetical job than did married women.[11] Spatial relations mediate the links between gender, family obligations and employment opportunities, and the spatial links seem most constricted for married women.

BREAKING DOWN STEREOTYPES ABOUT MARRIED WOMEN: POSTMODERN 'NUCLEAR' FAMILIES

We have emphasized the differences between women who head households and those who live with men. It is misleading, however, to treat 'married women' as a stable, unified category. A static reading of 'married women' also has practical implications that we find unacceptable, aligned, as it seems to be, with a separatist politics and pessimism about the possibilities of social change within heterosexual families. Ironically, a static reading of the nuclear family downgrades the real impact that feminist political action has had on family life over the last twenty years (Stacey 1990).

Heterosexual couples in Worcester choose many different ways of organizing domestic life and work. We attempted to capture these variations by asking about the division of domestic work within the household: who, for example, cooks the meals, washes the dishes, cleans the house? As a rough way of quantifying a woman's domestic workload, we tabulated an index of household responsibility. A woman received a point for each of the following: the presence of children under six years; if she does all of the house cleaning; has sole responsibility for cooking the meals; and does most of the cleaning up after meals. Table 2.6 shows that there is a real range of responsibility for domestic work among employed married women and that this bears upon their paid employment. Women who live in households in which domestic work is more equitably distributed between men and women are more likely to work in gender-integrated or gender-atypical occupations. Although the direction of causation between domestic workload and type of occupation is difficult to disentangle (i.e. does a job in a male-dominated occupation precede or follow a more equitable distribution of tasks within the household?), this finding brings us back to Siltanen's point: it is not gender or even marital status, *per se*, that is significant for understanding occupational segregation but, rather, a more specific set of domestic relationships. (But this generalization is complicated by our finding that employers do seem to operate with a reified set of marital status categories.)

Table 2.6 Household responsibilities of married women employed in different occupational types

| | Occupational type | | |
Index score	Female-dominated	Gender-integrated	Male-dominated
0	3.9%	6.2%	11.1%
1	15.7	27.8	33.3
2	20.3	29.9	16.7
3	43.1	25.8	33.3
4	17.0	10.3	5.6
N =	153	97	18
X^2=17.89, d.f.=8, p.=0.02			

Index score ranges from 0 to 4. Individual gets a point for each of the following: has children under six years old; does all cleaning of the house; cooks all the meals; cleans up after meals.
Source: Personal interviews, Worcester, Massachusetts, MSA, 1987.

One must be very cautious, however, in extrapolating the effects of sharing domestic responsibilities to occupational segregation. We have found that many working-class Worcester parents share childcare by arranging paid employment sequentially: almost one out of three dual-earner households with children under thirteen made these arrangements (Pratt and Hanson 1991b). While domestic responsibilities are being shared to some extent, we found that for two-thirds of the households using this childcare strategy, it was the woman who took the less optimal or less conventional time slot, usually a second (4.00 p.m. to 12.00 p.m.) or third (11.00 p.m. to 7.00 a.m.) shift, and this tended to drive these women into female-dominated jobs available at these times (e.g. nurse's aide, word processing and data entry). Sixty-five per cent of women in households that arranged paid employment sequentially worked in female-dominated occupations, compared with 51 per cent of women in households with no sequential shifts.[12]

The effects of a sequential shift strategy on women's labour force participation are likely to be highly contingent on the local labour market. In her study of Portuguese and Colombian immigrant families living in Central Falls, Rhode Island, Louis Lamphere (1987) documents the prevalence of the sequential shift strategy. In this context, however, the majority of men worked the second or third shift (and their wives the first), most likely reflecting the male's immigrant status and type of work, low-skilled factory jobs, mostly in the textile industry and most readily available to immigrant males at night. Even within the Worcester metropolitan region, the local opportunities that women and men have for obtaining part-time and/or shift work vary considerably by local area, as shown in Table 2.7. In the Blackstone Valley almost one out of every four employers interviewed hired women for evening or night shifts. In the inner city of Worcester, only one in ten firms hired women for the evening or night shift. These spatial variations in employment opportunities, paired with the tendency for women to work close to home, suggest local variations in the organization of work within households and, in turn, the construction of gender relations.

Table 2.7 Opportunities for part-time and shift work in manufacturing and producer services firms in four local areas of Worcester

	Area			
	Westborough (middle class, suburban) N = 44	Main South (working class, inner city) N = 49	Upper Burncoat (mixed class, outer city) N = 16	Blackstone Valley (working class, suburban) N = 31
% firms offering part-time work	70	51	56	81
% firms hiring women part-time	70	41	50	52
% firms that have evening or night shift work	27	25	50	50
% firms that hire women for evening or night shift	11	8	19	24

Source: Interviews with Worcester Employers, 1989.

CHANGE THROUGH THE LIFE COURSE: MIDDLE-AGED MARRIED WOMEN

Our Worcester study illuminates the impossibility of writing a single script for all married women. A close look at the lives of middle-aged married women reveals a diversity of paths and of changing relationships to the formal labour market over time. As many employers and much of feminist theory would have it, women's life-long relationship to the labour force is dictated by their involvement in marriage, child-bearing and child-rearing. To trace the actual impacts of these events on the later stages of women's lives, we looked at the women in our sample who were no longer actively involved in child-rearing. All of these women were over forty (but under sixty-five), were currently married and had never been divorced, and had no children under eighteen living at home (although many had children in their late teens and twenties, and even thirties, living with them!).

A close reading of the interviews with these women, whom we shall refer to here as the mature married group, reveals their keen awareness of the greatly expanded opportunities now available to girls and women compared to the options they perceived for themselves when they were growing up. In reflecting on how they came to do the type of work they do, and on their home–work priorities, nearly all of them pointed to the constraining expectations placed on them as girls: to work (if at all) as secretary or nurse or elementary school teacher, but primarily to devote themselves to a husband and children. A fifty-four-year-old woman who currently holds down two part-time teaching jobs said:

> 'When I got married, I thought experience as a teacher would be good so that if I ever had to go to work, the hours would be similar to the kids' hours. My husband and I saw it [my training to be a teacher] as a good insurance policy. Then when I had the kids, I never seemed to have time to work. I just missed the women's movement. I felt the effects of it, but I wasn't as motivated to find a career as people five years younger than me are. If I were five years younger, I probably would have worked full-time. My husband was not anxious to have me work, so for me to take a part-time job was more reasonable for us. Way back then it seemed that being a teacher was a good thing for a woman with a family. Back then I asked "What careers will fit a family?", not the other way around.'

Another woman, sixty-one, who works full-time as a secretary at a local high school said,

> 'In my day you had a choice of secretarial or home economics as a course of study. I was in the era when career possibilities were not

as open to women as they are today. The first responsibility was to home and the job was second.'

Another woman, sixty, who now works part-time as a school psychologist summed up her view of how the expectations placed on her generation met with the realities of social change:

'I'm in the transitional generation of women which is neither fish nor fowl.'

Clearly, these women were expected to follow a life course script that was very much in line with the 'essentialist' position we sketched out earlier. To what extent have their lives followed the prescribed script?

While almost all of the 138 mature married women in our sample did in fact leave the labour force in order to devote some time exclusively to marriage and child-bearing/rearing, only four of them (less than 3 per cent) never returned to the labour force after having children. Most (78 per cent) are currently working. A comparison of this mature married group with other married women reveals some interesting, if perhaps predictable, changes in women's relationship to the paid labour force as child-rearing demands subside. First, the mature married women who are employed outside the home are more likely to work full-time (as opposed to part-time) than are other married women,[13] and most (about two-thirds) of both groups of full-time workers say that they work full-time because they want to – that is, they would not prefer to be working part-time. The mature married women who do work part-time are, however, more firmly devoted than are the other married women to part-time work: not one part-time worker in the mature group said she preferred full-time work, whereas 15 per cent of the other married part-time women said they would rather be working full-time than part-time.[14] The mature married women appear, therefore, to be better able than the other married women to pursue their preferred level of labour force participation, with one group intentionally working only part-time and the other full-time.

A second shift in women's relationship to employment over time is evident in their evaluation of job attributes. Mature married women are, quite predictably, less constrained by the need to fit work around the demands of childcare (Table 2.8). When asked about the importance of different job attributes in general, the mature married women were significantly less likely than were the other married women to place great importance on the job hours when they had to be at work, on having flexible job hours or on the fit between the job hours and childcare or school schedules. Instead, a higher proportion of the mature group than of the other group said that the amount of independence on the job was important or very important.

Table 2.8 Married women's evaluation of job attributes in general

	% Saying attribute is important or very important		
	Mature	Other	p-value*
Close to home	85	79	
Easy to get to	86	78	
Low transportation expense	62	56	
Proximity to childcare/schools	58	72	0.01
The job hours	72	86	0.01
Having flexible hours	63	80	0.01
Possibilities for advancement	61	72	0.05
People you work with	91	87	
The type of work	94	93	
Good pay	88	92	
Good benefits	85	76	
Amount of independence	92	80	0.02
Amount of prestige	47	41	
Physical work environment	70	67	
Fit with partner's work schedule	79	81	
Fit with childcare/school	63	88	0.01

* p-value resulting from X^2 test contrasting the two groups' evaluation of each attribute as either (1) very important or important; (2) neither important nor unimportant; and (3) unimportant or very unimportant.

The way in which mature married women evaluate jobs in general is reflected in the actual nature of their present employment. When asked how well each of the phrases in Table 2.8 described their present jobs, a significantly smaller proportion ($p<0.10$) of mature married women than of the other married women said that their jobs were located close to childcare or schools, had flexible hours or had hours that fitted with childcare or school schedules. Clearly, one particular set of space–time constraints surrounding the job choice of employed married women with childcare responsibilities has been relaxed for the women whose children are grown. The reasons women gave for choosing their present jobs suggest that the mature married women find it easier than do other married women to pursue the line of work they have been trained for or the type of work they want to do: 46 per cent of the mature group versus 33 per cent of the other group said that they took their present jobs because it was the type of work they had been trained for or the kind of work they wanted to do. Without the constraints of having to select a job on the basis of its hours, its flexibility and its proximity to schools/childcare, the mature women seem to have more freedom to pursue a type of work that interested them. They are no more likely than are other married women, however, to have jobs in gender-atypical occupations.

Although some of the space–time constraints facing employed women in the child-rearing years no longer apply for the mature married women, the major spatial constraint (the importance attributed to having a job close to home or one that is easy to get to) is not at all diminished for this group of middle-aged women; the location of the job close to home remains a highly valued job attribute (Table 2.8). That having a job close to home remains important, even after a woman's children are grown, is evident in the fact that the mature married women in our sample had journeys to work that were no longer than those of other married women (both averaged about sixteen minutes). Nor were the mature married women any more willing to consider the possibility of taking jobs that were farther from home than were the younger married women (for both groups the maximum commute they could envisage was about twenty-seven minutes on average).

The sustained importance of location probably reflects the fact that many of the mature married women see their paid jobs as secondary to their primary job, which remains in the home. This is suggested by the fact that relatively few mature married women were concerned with a job's possibilities for advancement (Table 2.8) or could envisage a promotion for themselves.[15] Some explicitly stated that they did not want their employment to interfere with their home life or their home responsibilities. For example, the school psychologist mentioned earlier said, 'I don't want my job to disrupt my homelife routine, the times when I need to be here.'

Another woman, aged fifty-six, who works part-time as a collator for a local newspaper, has worked part-time on and off for many years and describes herself as 'basically a housewife and a mother and I find nothing wrong with that at all'. At the same time, many mature women saw their jobs as providing the household with an added dimension of financial stability and security, should their husbands be laid off.

The persistent tug of home, and therefore the reluctance to incur a long commute, is evident in the stories these women tell about how they approached the labour market after an often extended spell at home. Re-entry to the labour force is often seen by the women themselves as unintentional, accidental and precipitated by the proximity of a suitable job opportunity, which the woman learns about serendipitously through family, friends, neighbours or the geography of her everyday life. A fifty-three-year-old woman who works full-time as a receptionist for a large computer firm, talked about her return to the labour force after an extended break during which she raised her eight children:

'This job just happened. The job became available, and it was just down the street. A neighbour who was working there asked me if I wanted the job, and I said sure, and I went down and applied, and I was accepted. I was in the mood for getting a job, but I wasn't really looking for anything. I was at an age when the kids were old enough for me to work, but I wasn't sure if I wanted to work eight hours a day, five days a week. My family kept encouraging me, so I did it. When I first took this job, the children were in junior high and high school. I wanted something close to home to be available to them. Also, I worked around the kids' schedule. I could be out of work early and be home to fix dinner.'

Another woman, now fifty-nine, recalls how she started back to work part-time for a valve manufacturer fourteen years ago after several years at home with her four children: 'My daughter went to [this employer] to find a summer job; they said they wanted someone on a regular basis. It was only five minutes from home, so I took the job. They trained me and it worked out well.'

Even when the decision to return to work is taken more definitively, mature married women often approach the labour market hesitantly, choosing to ease back into paid work through taking a part-time job or a job with a temporary agency. A fifty-six-year-old data base coordinator with a large computer firm admitted that after being out of the work force for twenty-three years she approached the labour market timidly: 'I started out as a temp for [this firm] and ended up staying. I figured temping would be the best way to get back into the work force – to work a few days a week and see if I liked it.' She had been working at this firm for eight years when we interviewed her and noted that the reason she works there is that 'it's convenient, it's close by, and because the job was offered to me – I didn't have to look for it'. A registered nurse aged fifty-one knew that she wanted to go back to work after five years at home caring for two young sons, but found she had to be coaxed back by a former co-worker:

'I thought I wanted to stay home with the kids, but after a few years of that I had to get out of the house. The administrator here [the hospital where she now works] knew me from another hospital, and she kept phoning me until I took this job. I was still nursing a new baby, but she persisted. First I went [to work] just every other weekend, when my husband was at home and could take care of the kids. Gradually, I worked more and more hours until eventually I went full-time.'

Most of the mature married women returned to work when they still had school-aged children at home and therefore only considered jobs that would permit them to combine paid work with their domestic responsibilities. A fifty-year-old woman who works as an accountant underlines what this means in terms of job location and job hours:

46

'When my oldest son was thirteen, I went back to work. I went to Wonder Market because it was part-time, convenient, close to home, and the children were at school. My work hours were perfect for their school hours, and I could get home in an emergency. I worked because I wanted to get out of the house. It was the perfect job for me because it was close to home, and I got all my accounting training on the job.'

We have seen, then, that with children's growing independence, married women are more likely to work full-time and to consider a wider range of job possibilities; yet the centripetal force of home on the daily orbits of most of these women endures. A closer look at the home/work connections of the mature married women reveals the many ways in which changing domestic responsibilities prompt changes in women's relationship to the paid labour market after children are grown. The household responsibility index described earlier shows a significant shift in the domestic workload as women move through the life course. Higher proportions of the mature married women, relative to the other married women, have both very low and very high values on this measure, indicating the presence of two groups of mature married women, one with quite heavy and one with relatively lighter domestic workloads (Table 2.9, left panel). Furthermore, within this older group, the level of household responsibility is inversely related to level of labour force participation: mature married women with greater domestic

Table 2.9 Household responsibilities of married women

Household responsibility index[a]	Mature married women[b]	Other married women	Mature married women who work full-time	part-time
0	27	21	23	3
	(19.6)[c]	(8.5)	(32.4)	(10.7)
1	23	25	15	2
	(16.7)	(10.5)	(21.1)	(7.1)
2	28	51	11	4
	(20.3)	(20.7)	(15.5)	(14.3)
3	60	78	22	19
	(43.5)	(31.7)	(31.0)	(67.9)
4	0	71	0	0
	0	(28.6)	0	0
Totals	138	246	71	28
	(100.0)	(100.0)	(100.0)	(100.00)

$X^2 = 54.8$ $X^2 = 12.5$
p=0.00   p=0.00

a See Table 2.6.
b Over age forty, never divorced, no children under age eighteen at home.
c Column percentage.

demands are significantly more likely to work part-time, and the mature married women with lighter domestic workloads are more likely to work full-time (Table 2.9, right panel). (This same relationship holds for the other married women as well.) We see, then, within this group of married women with grown children, considerable variability in level of domestic responsibility and the mirroring of this variability in level of labour force participation.

Moreover, as women move through the life course, they face new kinds of family responsibilities not acknowledged in an essentialist view that fastens on the child-bearing and early child-rearing years. The mature married women in our sample, while not responsible for young children, were more likely than were the other married women to have someone else – other than children – who relied on them for care on a regular basis; about one-third (32 per cent) of the mature married women, versus one-fifth of the other married women, said that someone, like a parent or a grandchild, depended on them for care on a regular basis. Many women in our sample found themselves, after their children were grown, altering their labour force participation to accommodate these new family demands. The accountant we described earlier, for example, was planning to shift from full-time to part-time work so that she could spend more time with her parents:

'By working full time I am too tired to see them very much, and I feel guilty. I want to be able to spend more time with them and also have more time around the house.'

Many other mature married women had switched from full-time to part-time paid work, or had left the labour market entirely to care for grandchildren or for spouse or parents. In other ways, too, the changing demographics of the family life course lead women to change their relationship to paid employment. Prime among these factors are the financial demands that a home mortgage or college tuition impose on a household. Re-entry to the labour force for many of the mature women in our sample had been spurred by the desire or the need to support a home mortgage or the children's college education.

A final theme that emerges from the interviews with the mature married women concerns the variety of experiences these women encountered after they rejoined the labour force. At least three trajectories are evident. Some women are able to pick up where they left off in the labour market and continue in much the same vein as before they stopped working. Others find their labour market skills outdated when they wish to return to work and end up taking less desirable jobs than those they had previously. Still others find the post-child-rearing stage exhilarating, one in which they can devote their energies to career growth. We offer a few vignettes to capture the flavour of each of these post re-entry trajectories.

48

The most familiar type of 'downward mobility' entails women who worked in office jobs before leaving the labour market, find their office skills unappreciated when they approach the labour market after a break from it, and end up in unskilled manual jobs such as janitorial or factory work. Alice, for example, is a fifty-five-year-old high school graduate who worked in an office before her children were born. In 1975, at the age of forty-three, she resolved to go back to work because she decided, 'the kids were old enough for me to leave them'. Alice wanted to return to an office job but was not able to find one. Her story illustrates the problems facing many middle-aged women wanting to return to the labour force:

'I looked for an office job but I couldn't get one. I only had high school, and they didn't want old ladies. So I took a full-time job as a housekeeper at City Hospital. After three years at that job, the person I car-pooled with had a heart attack, and I couldn't get to City without taking two buses, so I got a new job working as a housekeeper in a nursing home because I could walk there. [She worked there for a couple of years, and then a co-worker at the nursing home told her about a job selling tickets at a local ice-skating rink.] It was a cleaner job and not as depressing, so for four years I stayed at home working only winters selling tickets at the ice-skating rink. It was getting hard with the car, though. We don't have money for two cars, and it got to be too much of a hassle. I had to wait half an hour for him [her husband] to pick me up. It was too late to make supper, and I was tired of it. I couldn't get back and forth [to work] the way I wanted to, so in 1984 I took a part-time job preparing the salad buffet at [a local supermarket]. I would drive my husband to work, then go to work, then pick him up and go home. But this job was only three-and-a-half hours a day and they wouldn't give me more hours, so I took a job as a bus monitor for the Worcester Public Schools full-time. It was more hours than [the supermarket] and it was convenient because I could walk. I left that job after a year mainly because the office moved and I couldn't walk to work anymore; I would have had to take two buses many times a day because of the split shift. I would have had to spend all of the money I earned on buses, so it was cheaper to stay home. So, since 1986, I have chosen to stay home. I can get my own work done, have supper ready when my husband and kids come home, and I am not tired.'

Alice's story is typical of the many mature women whose work after labour force re-entry does not match their earlier work. Although Alice does drive, the combination of not being able to find the type of work she wanted to do and persistent transportation problems led her to take a series of dead-end jobs and eventually to leave the labour force entirely.

Rita, aged fifty-two, sees her downward trajectory in terms of the clash between home and work roles.

'I started out as a secretary, got married, had kids, and when I decided to go back to work twelve years ago, I wanted mothers' hours so that I could take time off to be with the kids. I wouldn't have gone into factory work, I just did it because the hours were ideal. In an office job, you can't be taking the summers off whenever you want.'

Although Rita had worked full-time for twelve years at an unskilled job in a plastics factory, she recently took a six-month accounting course and hopes to find a bookkeeping job. Rita's case points to the reversibility of a downward trajectory, contingent on demands at home.

Doris, aged fifty-one, is an example of the group of mature woman who were able to return to the same type of work they had been doing before they left the labour force. Doris had done office work before she was married and has been working full-time in data entry at a large bank for the past fifteen years.

'I had been working in schools and newspaper offices – little part-time jobs. My neighbour started to work at [this bank]. My kids were in school, and when she told me about this job, I took it out of convenience and the necessity to work.'

Later in the interview she echoes a theme that resonates throughout almost all of the mature married women's interviews:

'Yes, I see myself as a housewife and as a mother first. My job is secondary; it's not the most important thing. I'm not that concerned with my career. I'd like to spend more time with my children and my grandchildren. I think women can do anything they want to today. For example, if I wanted to go into nursing, I could. If I really wanted to go back to school and do what I wanted, it's available to me. But at my age, I'm more concerned with retirement.'

Other mature married women may have returned to the labour force simply to get out of the house or to earn needed cash, but then they find that with reduced child-rearing demands, they can devote more of their energy to their careers. Sharon, aged forty-one, who works as a food service supervisor at a nursing home illustrates this trajectory.

'I started back to work twelve years ago working in a hotel kitchen part-time for a friend. More or less, I went to work for something to do, so I worked when they needed me and when I wanted to.

50

It was very flexible. My son was older and my husband was working, and I filled a gap in my life by working. Then I went to [a nursing home] and worked under the direction of someone else for four years. In my third year there, I decided I wanted to be a supervisor, so I got a food service certificate. I went to school, and then I saw an ad in the paper for this job and I got it. [Although Sharon does not want to be the family breadwinner, career advancement is clearly important to her now.] I don't want to be stuck in one place. I go to seminars now and then, and I took a management course last year, and I'd like to take a course in computers because a lot of kitchens are using computers now and it would help me to move up.'

Patricia, aged fifty, also exemplifies mature women whose career interests increase as child-rearing demands wane. She now works full-time as an occupational nurse at a large computer firm:

'I was a night nurse at [a local college] but there was no growth there. I didn't see any changes, it was boring. It was depressing working nights, but I wanted to be home with the kids during the day, and I had not worked for eleven years. So it was a good job for that. I had a friend who had told me about an opening there, I applied, and got the job. Now it has become more important to me to have growth in my job, to make more of a commitment to it. My family doesn't need me as much any more. Now I need something interesting outside. As the kids move through school, I'm going to wind down. I'm looking forward to retirement. I want to develop interests that I don't have time for now.'

Roberta, also aged fifty, talks of being more committed to her job now that her children are older. She has worked for the past eleven years as a secretary/administrative assistant at the Catholic Youth Services Center. She says,

'I guess I used to be concerned with making a fairly respectable salary. Now since the kids are grown – this really isn't a flash-in-the-pan job, and I really want to do a good job.'

Roberta exemplifies the group of women who come to see an old job through new eyes as home circumstances change.

These vignettes highlight the diversity of middle-aged married women's circumstances; there is no unidirectional course of individual downward job mobility. This is certainly the trajectory for some and we do see the commitment to home and family reverberating through the later years of most women's lives. Nevertheless, despite the fact that this cohort was subjected to strong expectations about their commitment to the home and the appropriateness of certain occupations for women, there are many women within it who have made new lives for themselves as their children age.

The sociological categories typically employed to measure occupational success seem woefully inadequate to the task of analysing these changes. In the last three vignettes outlined, the cases of Sharon, Patricia and Roberta, none of the women move out of 'female-dominated' occupations, educational upgrading takes the form of incremental training on the job that is difficult to capture through traditional credentials, and they do not obtain the professional/managerial status that figures prominently in liberal feminist 'success stories'. And yet their relationship to their jobs has changed: they are more committed to their jobs, and want and seem to experience change and growth in them. Patricia signals another change that is difficult to quantify and assess in relation to a script that measures success in terms of occupational attainment: the importance and pleasures of interests outside of the labour market, ones that she eagerly anticipates developing after retiring from paid employment. As the ongoing tension within feminism (between discourses that stress equality and those that highlight difference) signals, we run the danger of writing women's lives according to a male script that privileges paid employment. In learning to write about women's lives, we must hold that tension in balance, so as to encourage the rewriting of women's and men's lives to enable equality within workplaces and the transformation of the meaning of work.

CONCLUSIONS

We have attempted to unravel some of the complexity of women's participation in the labour force by reading diversity of experience through the lens of personal history and place. As Joy Parr (1990: 10) notes: 'there are many ways of being in gender within a single [historical] time and space.' The experience of being a woman or man differs depending on time in the life course and the community (even within a single metropolitan area) in which one lives.

We have fixed married women's labour force participation in opposition to that of divorced, separated and widowed women, arguing that the former are more likely to take up traditionally 'female' paid employment because of their extensive domestic responsibilities. In doing this, we have no doubt suppressed some continuities and silenced other differences, such as those based in class or race. In recognition of the danger of reifying a new set of categories based on marital status, however, we have attempted to 'open up' the category 'married woman' by exploring the diversity of married women's experiences of domestic work and labour force participation and how the experience of one is linked to the other. We have also explored how a married woman's work changes through the life course, noting that middle-aged women are more likely to work full-time outside of the home and to value independence in their job.

These analyses help to unfasten an essentialist link between gender and work experience. At the same time, the specificity of the links between unequal domestic responsibilities in the home and occupational segregation strengthens and sharpens a feminist critique of 'traditional' arrangements of domestic work. We were genuinely shocked to find that in Worcester it is almost unheard of for women to combine mothering work with a well-paid job. It is also apparent that most mature married women had few illusions about career advancement and few had metamorphosed into successful career (paid or otherwise) women. The spatial constraints that tie younger married women to extremely local labour markets continue to structure the employment choices of older women. Many middle-aged women with whom we spoke placed themselves in history as a 'lost generation' of women. It is an open question, however, as to whether their perceptions are correct; we see little evidence that the majority of the next generation of married women will have very different lives in their middle ages.

The answer to this question about social change is unlikely to be consistent across places. The absence of younger 'super women' in Worcester may reflect aspects of local cultures, local familial expectations, local employment opportunities and local amenities for childcare. Even within the Worcester region, we have argued, there are very different opportunities for family members to combine paid and domestic work, depending on the mix and timing of local employment opportunities. This suggests that we must be attuned not only to differences across the life course, but to the varying ways that similar life paths are structured in different places. Place, as a further source of difference among women, deserves fuller attention. At the same time, a life course approach lessens the possibility of reifying the links between social and spatial relations. Numerous geographers have drawn attention to women's tendency to find employment close to home and the ways that dependence on very local job opportunities limits women's occupational choices. Although we do find that mature married women continue to search for and value paid employment close to home, there is some evidence that divorced women are willing to search further for paid employment. In other words, the constraints of space are felt differently by married and divorced women.

We have used the labour force experiences of divorced, separated and widowed women to sharpen our critique of the nuclear family, but a close look at their experiences extends a feminist critique in new directions. On the one hand, many divorced women are more successful in obtaining jobs in gender-atypical occupations and attain higher incomes than married women, often as the result of ambitious efforts to re-educate themselves in their adult years. On the other hand, the household incomes of many divorced women are dismally low relative to couple-headed households. Judith Stacey (1990: 45) remarks on this contradiction with reference to the women who she studied in the Silicon Valley: 'Heady new ideas, skills, self-esteem, and goals, each begun [in conjunction with divorce] a journey up the ladder of educational and occupational status, a journey that paradoxically hurtled each woman and her children precipitously down the economic pyramid.' The relative poverty of households headed by divorced women no doubt reflects the demise of 'the family wage'. The decline in real wages and the increased hours of paid employment required to sustain households may represent a point of potential alliance between divorced women and couple-headed households.

By loosening the links between gender and particular work experiences, we suggest the possibilities for social change. By carefully studying differences among women, we isolate more accurately the processes that link the gender division of labour within the home and labour market and some of the points of common interest between different women and men. Writing about these points of difference and commonality is one part of the process of rescripting women's lives.

3

ELIMINATING THE JOURNEY TO WORK

Home-based work across the life course of women in the United States

Kathleen Christensen

INTRODUCTION

Most of the geographic literature on how women combine their productive and reproductive lives has focused on how they manage the journey to work (Cichocki 1980; Coutras and Fagnani 1978; Ericksen 1977; Fagnani this volume; Kaniss and Robins 1974; Madden and White 1978; Palm and Pred 1974; Pratt and Hanson this volume). This chapter explores another geographic option, that of bringing paid labour into the home. By converging work and family in one place, women eliminate the journey to work, but create entirely new issues and problems that vary according to their stage in life. Through the use of case studies selected from my research, I want to examine why women work in their homes and how the experiences of working there relate to the larger conditions of their lives.

Research background

In January of 1985, I conducted a national survey of 14,000 women in the United States, half of whom wanted to work in their homes, half of whom actually did. Of the approximately 7,000 women who worked at home, the majority were white, married, and part-time workers, working an average of twenty hours a week. Nearly 80 per cent of them were self-employed and worked alone at home (Hirshey 1985: 70; Christensen 1988). Most of the women contributed supplementary income to their households, with 64 per cent providing less than a quarter of the total income.

Half of the women had children under eighteen years of age; half lived in childless homes, either because their children had grown and moved away, or because they had no children.

55

Subsequent to this 1985 survey, I conducted in-person and in-depth interviews with 100 of the women who had responded that they worked at home. The women lived in predominantly middle-class suburbs of New York, Chicago, and San Francisco. They worked in a variety of skilled occupations, including clerical, technical, and professional. And as representative of the survey sample, half of them had children under the age of eighteen and half did not.

For the purposes of this chapter, I selected case studies of four women, two of whom were at the child-bearing and -rearing stages of their life course, two of whom were not. I intend to let the four women speak for themselves on their own terms and in their own words. Prior to presenting the cases, however, I would like to orient the reader to some of the life course themes that their stories reveal. In the concluding sections of the chapter, I shall address the consequences of working at home, including the effects on the micro-scale geographies of the home and on the health and pension protection for women who work at home, as well as the policy directions that the public and private sectors could pursue to ensure that working at home does not penalize women.

LIFE COURSE THEMES

In looking at the individual case studies, three broad themes are clear: the effects of a woman's generation on her attitudes and decisions; the effects of a woman's prior personal history on her subsequent job decisions; and the psychological transformations a woman can undergo as she moves through her life.

Effects of one's generation

The women presented in the subsequent case studies fall into two generations: those born in the 1930s, who bore and reared their children in the 1950s; and those born in the 1950s, who started to have their children in the 1970s. Each generation was shaped by the prevailing norms of their time regarding women's roles in the home and the workplace.

The expectations for women in the older generation were quite clear. They were to be first and foremost wives, mothers, and homemakers. If they went to college, it was almost always to be trained in something like teaching or nursing, upon which they could fall back if their husbands died, were disabled, or lost their jobs. If they were to work for pay, they planned to do it before their children were born or after they were raised.

The economics of the 1950s in the United States made it possible for women in these white, middle-class households to live this ideal. The income of the father could easily support a comfortable lifestyle for his family, including the purchase of their own home. Popular culture through television series, such as 'I Love Lucy', 'The Donna Reed Show', and 'Leave It to Beaver', reified this notion of the breadwinning father and the breadmaking mother. Although many women in this generation were entirely comfortable with these role expectations, others found them suffocating. Perhaps no one expressed this discontent more clearly than did Betty Friedan in her 1963 classic, *The Feminine Mystique*, which helped trigger the subsequent development of the women's movement.

By the 1970s many of these older women were trying to re-enter the labour market. Some did so for economic reasons; the recessions of the 1970s forced them back into wage-earning work. Others did so for want of other activities; often after their youngest child left for college, they decided to pursue a job as a way of keeping busy and providing an outlet for their interests. And still others were profoundly influenced by the women's movement and felt that they deserved to create careers for themselves rather than simply supporting the lives of their husbands.

More often than not, these homemakers of the 1950s found that potential employers did not recognize, let alone place a high premium, on the skills they had developed during their years away from the workplace. As a result, this older generation of women often were forced to take jobs that were well below the skill levels they had developed prior to the birth of their first child.

The younger generation of women were brought up in an era in which the messages and expectations for women were more mixed than they had been for their mothers during the 1950s. On the one hand, they were born during the heyday of the glorified wife and mother, and naturally had internalized some of these cultural norms. On the other hand, they came of age during the 1960s, when that glorification was under assault by the women's movement, which placed a high premium on college education, careers, and eroding the sexism that pervaded the home and workplace. The operative word was 'opportunity', and a popular slogan of the time was 'A woman's place is in the house – and Senate.' Popular culture extolled this new woman in television shows such as 'That Girl' and 'The Mary Tyler Moore Show', both of which portrayed positively the young single woman pursuing her career.

Women of the 1970s were not uniform in their acceptance of the new woman. Some chose to retain the traditional ideals of full-time wife, mother, and homemaker, while others sought education and careers, typically delaying marriage and motherhood.

Younger women with more traditional values, however, have faced more difficulties than did their mothers in trying to live out their ideals of full-time homemaker and mother. The combination of recessions, lay-offs, and inflation throughout the 1970s eroded job security, making it difficult for many families to depend on just one wage to support the family. Real income has fallen steadily in the US since the 1970s, making it financially impossible for most households to live a middle-class lifestyle on a single income. Even with dual wage earners in the majority of families, the percentage of home ownership in this country is at the lowest level it has been in the last several decades (*New York Times*, 1991: 18). As a result, high school educated women who assumed that they would work only until they married and gave birth, have now found themselves looking for jobs in a market-place that increasingly expects college-educated workers. Not only are many of these women unhappy to find themselves as 'working mothers', they are distressed to discover that decisions made earlier in their lives regarding education are proving disadvantageous to their potential earning power.

Women who accepted the messages of the women's movement find themselves in difficult positions in the 1990s, for reasons different than those faced by more traditional women. These feminist-influenced women took seriously the exhortation to get an education and build a career, often postponing marriage and children. Many of these career women now find themselves in their thirties and forties wanting children and trying to conceive, while their 'biological clocks' tick inexorably more slowly. The threat these educated career women heard repeatedly during the late 1980s was that a single woman in her thirties was statistically more likely to be the victim of a terrorist's assault than to get married.

These concerns about the limited possibilities of having a satisfactory personal family life are central to many career women. When they do have children, they experience a great deal of conflict: on the one hand, they want to spend time with their children; on the other hand, they are unready to give up a career that they have spent over a decade building. It is not surprising, therefore, that when older women have children they consciously seek to find a harmonious balance between their productive and reproductive lives. Although men married to these women have often increased their roles in the domestic labour of the home, the majority of the emotional and practical work continues to be assumed by women (Hochschild 1989), partly because men have not accepted an equal role in child-rearing and housework and partly because of the persistence of the notion of what constitutes a 'good mother'.

Many women, as well as men, still carry in the recesses of their minds a specific notion of the 'good mother' that was shaped profoundly by their own experiences of being raised by 1950s women who stayed at home. A good mother is one who spends a great deal of time with her children, teaching them the differences between right and wrong, feeding them, bathing them, and playing with them. The possibility of ceding these activities full-time to nannies or other daycare providers remains a foreign notion that to many women violates the code of the good mother. The notion of the good mother as the stay-at-home mother is prompting many career-oriented women to pressure employers to provide flexible work options, such as part-time work and job-sharing for parents (Christensen 1989).

It was within these broad social changes of the 1950s and 1970s that women have had to make individual decisions regarding work, education, and motherhood.

Effects of prior personal history

Throughout my research it became clear that the women I interviewed fell into one of two groups regarding individual choices about education and work. Although heavily influenced by the broad social climates of the time, most saw their personal decisions as individual ones rather than generational ones. There were those who graduated from high school and sought jobs, and those who went to college and expected to have careers. Their work-related aspirations corresponded to their broader self-definitions as 'traditional' versus 'career-oriented' women.

Job-seekers typically expected to work only until they married and had their children. They saw their work as a finite set of activities for which they received a set amount of money. There was minimal personal investment in the work or identification with it. Career-seekers, on the other hand, saw college and post-degree education as the avenue they needed to take to establish their careers. Once in a job, they identified strongly with their work and they expected a range of pay-offs from their work, including social, psychological, and remunerative rewards. They viewed each position as a step on a career path.

Individual women often found themselves with changed aspirations during their life course. This was more often the case with traditional women who decided they wanted or needed more of a career than what was offered them by the educational decisions they had made earlier in their lives. As a result, their degree of freedom in achieving their new work aspirations was constrained.

As will be clear in the subsequent case studies, a general principle holds true: decisions made early in life shape the range of possible decisions that can be made later. Efforts to override earlier decisions entail significant psychological transformations in how a woman thinks of herself as an agent.

Psychological transformations

Although heavily influenced by their generation and by their early life decisions, women remain active agents in their lives, exercising a measure of freedom to alter their life course and options. The ability to transform oneself becomes particularly apparent in women after their children are grown.

My research indicates that women in their forties and fifties typically feel freer than younger women to shake off others' expectations of themselves. Whether it is an expectation of a good mother, a solid worker, or a good wife, these women find themselves better able to define the conditions of their own lives. Making these changes in self-definition with regard to their work does not come without costs, however. Many older women I interviewed found their marriages weakened or destroyed as a result of their commitments to finding ways of expressing their own ambition, ability, and creativity.

Although the four women in the following case studies speak for themselves, each presents clear evidence of how their decisions regarding their productive and reproductive lives were influenced by the prevailing norms of their generation, as well as by individual decisions made earlier in their lives. In addition, the two women of the older generation also reveal the kinds of psychological transformations that are possible. In all of the cases, the decisions to work at home are tied to the broader conditions of their lives.

CASE STUDIES

Mothers of young children

More than 60 per cent of all mothers in the United States have paying jobs, but combining work and family proves difficult, especially when the children are young. According to the most recent US Bureau of Labor Statistics figures, for nearly 600,000 women with children under six years of age, a home-based job has been their attempt at a solution (Horvath 1986). For the traditional woman, working at home provides the closest alternative to being the full-time homemakers they want to be. For the more career-oriented ones, working at home provides a way to maintain a connection with their careers while raising their children.

To protect the identities of people described in this chapter, the names of the individuals have been changed. All quotations are real, however, and come directly from the interviews I had with the women or their husbands.

Anne Michado

A first-generation American, Anne was raised in a deeply religious Italian community in the Bronx, New York City, where the emphasis was on family and hard work. Although always industrious, Anne did

60

not have career aspirations when she graduated from high school. She took a clerk-typist job with a New York City bank with every intention of quitting as soon as she got married. Yet she began to enjoy her work and so she kept working after her wedding. Eventually she reached the highest rung of the secretarial ladder, looked higher, and saw that she could not go any further. After twelve years with the bank, she painfully realized that she would never be promoted out of the ranks of secretary. 'Here I was running this branch. I could do *all* the jobs, but I was constantly being passed over for promotions.' Bank management finally made it explicit that unless she went back to school and got her college degree, she would stay where she was.

By this time Anne was thirty. She had been married for five years and had worked at the bank long enough to be fully vested in the pension plan. She decided to forgo college – and a career – and to focus instead on her family. 'Sure I was a little bitter, but I had always wanted to have children, so I decided that this was the time to get pregnant.'

Anne and her husband assumed that once she gave birth, she would raise their child and not work for pay. But economic circumstances intervened. Her husband was temporarily laid off from his job as a building maintenance supervisor; she developed costly medical complications during her pregnancy; and an infant proved more expensive than they had expected: 'Money-wise, things got pretty rough.' Shortly after her daughter's birth, Anne decided she had to find a part-time job.

The reality of the part-time job market quickly turned her sour on the idea of working outside the home. Since she and her husband had no car, Anne took the best job she could find on their bus route: as a cashier earning minimum wage at a local fast-food store. In order to save baby-sitting expenses, she worked nights. As a result, she almost never saw her husband. He would walk in at five, just as she left for work. After several months, Anne decided to find a job she could do at home, scheduled around her nine-month-old's sleeping schedule. She happened on her job which was advertised in the local paper:

> 'Finally, I saw an ad for a home typist. They offered pick-up and delivery and it seemed perfect. I answered the ad, went down and took a typing test from a dictaphone, passed, and was hired. They explained that I would be paid fifty cents a page and two cents a line for anything over two pages. It ends up that I average about $5 an hour.'

This was over a dollar more than she had earned as a cashier. Out of these earnings she paid $65 a month rental for the typewriter.

Anne thought she was an employee of the typing service. Although the company told her they were not going to deduct taxes, Anne did not realize that meant she was being hired as a self-employed independent contractor, and it was not explained to her. She only realized the true nature of her employment status a year later at tax time, when her husband took their forms to a tax accountant. Then she began to compare the costs and benefits of this arrangement with those of being a company employee.

Anne is paid a piece-rate for each page she types. She gets paid only for the amount of work she does, not for any lag times between work, nor does she get any paid vacations or sick leave. She receives no health insurance or pension coverage, and because she is self-employed she pays more into her Federal social security account than she would as an employee. At the time I interviewed her, as a self-employed contractor she paid 12.3 per cent of her earnings for social security, whereas had she been an employee, she would have paid 7.15 per cent and her employer would have contributed the difference. Her employer also made no contributions to any other government protection for her. If she ever hurt herself doing her work, she would not be eligible for worker's compensation; if her company stopped giving her work, nor would she be eligible for unemployment insurance (Christensen 1986).

Anne knows that she made certain trade-offs by working as a home-based contractor but sees it as the best of not-very-good alternatives: 'I get low pay, boring work, no opportunities for advancement, and no benefits, but I get to stay home with my daughter.' And working at home is easier than working either part-time or full-time outside of it.

> 'I don't know if I could handle a full-time job outside. I'd have to leave the house at seven. Then I'd come home at six or seven and then I'd have to cook and take care of the family. I don't think I could do that. That is just more pressure on top of pressure.'

In my follow-up interviews, I found that three out of four home-based working women assumed overall responsibility for housework. The only task their husbands were likely to take on was dealing with the refuse. It is not surprising, then, that many women make the decision to work at home. A full-time job outside the home would be too demanding and a part-time job would often not pay enough to make it worthwhile. Given these options, Anne believes that, all things considered, working at home suits her needs at this time in her life.

Yet, the arrangement has its own stresses. In order to get the work finished, she keeps to a gruelling schedule. When she first started as a home-based typist Anne worked for an hour in the afternoon while her infant napped, then she prepared dinner, cleaned up, bathed the child and put her to bed, and sat down to work again at seven o'clock in the evening. Spurred by her deadlines, she would work until two or three o'clock in the morning, then get back up at six o'clock to see her husband off and care for her daughter. After six months of getting three or four hours of sleep a night, she quit.

Anne waited until her daughter was three years old and could go to a play school three mornings a week before she started typing again. She now types during the morning hours, and again in the evenings from seven o'clock until midnight or one o'clock in the morning. Having additional daytime hours has not eliminated her night work, but it has shortened it. She works in the evening while her husband watches their child. The two of them go upstairs to the family room and watch television or read. She finds it very hard some evenings to stay at her kitchen desk typing when she knows that they are curled up on the sofa watching TV.

Despite the mixed blessings of her arrangement, overall she likes having her typing job – it gives her something to do besides housework. 'There's only so much cooking and cleaning I can do.' Yet she is angry at what she feels are society's unrealistic demands on women. 'Just being a housewife is not supposed to be enough, you know? We're expected to take care of the family, run the house, and hold a job.'

Anne is angry about the way the advertising media portray housewives as stupid and working women as better. She particularly hated the TV ad several years ago that had a tall, leggy brunette crooning, 'I can bring home the bacon, fry it up in a pan, and never let him forget that he's a man.'

> 'I go "Goddamn it, will you listen to this?" They make it sound like you love doing all that, and you *don't*. It really bugs me; it just upsets me. I wish I could win the lottery and then we'd be comfortable, but I can't.'

The reality is that Anne no longer has the choice whether or not to work. She has been able to choose where to work, but feels that the difficulty of trying to separate work and family in the home makes it even harder than working outside.

> 'You're juggling so many things, and it's constantly there. When you're in an office, the only thing you really do is work. You might think of the house, and say, "Oh God, I've got to cook something for dinner", or "Did I defrost the roast?" But when you get up to go get a soda from the soda machine, that's all you're doing. At home I get up to get a soda, and the refrigerator says, "Better clean me." Or I go to the bathroom and I think, "I better clean that." It's always there, hitting me in the face.'

Since she is 'at home', Anne has in fact assumed much more of the responsibility for housework than she did when both she and her husband worked outside.

'The reason is simple. My husband goes to work, and I do not. I am home twenty-four hours a day and things kind of slip back to the age when the mother was always at home with the children. Slowly it turns back to that. I think even women begin to think like that. "It is more of a woman's job to take care of the house and family." We just slip back into it.'

Anne makes herself feel better about the toll working at home has taken on her by arguing that women are more adaptable than men:

'My husband could never do what I do. I don't think he could keep up the pace I do, getting only four or five hours of sleep every night for a year and a half. He needs more sleep than that; he conks out at ten or eleven and sleeps until six. We women can handle more. I think we are built differently. Maybe it is just because of the sense of family, but it makes us push harder. I think women adapt better than men. I just think women are more resilient.'

The notions of adaptability, strength, and resiliency all carry very positive meanings. Yet they obscure the reality that women like Anne are exhausted and getting limited help from their spouses. Although they may help out with the children, as Anne's husband willingly does, they do not pitch in with the cooking, cleaning, shopping, or worrying about getting it all done. Rather than confronting or criticizing their partners for being inflexible or unhelpful, the women make a virtue out of their own adaptability. Adaptability becomes a valued feminine quality. But in another way it becomes a survival technique, because making the family and marriage work is their primary goal. If their husbands are unwilling to confront or deal with the problems, the women have two options: force the issue with their husbands, at the risk of making everyone unhappy and possibly jeopardizing their marriages; or take on more of the responsibilities themselves and turn their ability to do so into a virtue.

At some point, the traditional women interviewed in my research tried the first option, but they all eventually tired of the struggle. They got worn down by having to fight on top of everything else – working, houseworking, and taking care of the children. Most end up with the second option, without realizing in the long run how destructive it is for them.

In contrast to the traditional homemakers, some women never intended to quit their jobs and 'stay home'. They had spent years advancing their careers, seeing each educational and job decision as a step in a sequence leading to greater and greater success. Their energies, their hearts, and their identities were heavily invested in their work. For the most part, these women married men who shared their belief that work and family do not preclude one other. It came as quite a surprise to many of these couples that their plans did not mesh with the realities of having a baby.

Although heralded as the first generation with choice – to work or not to work, to take time off to raise children or not – many of these career-oriented women say that they do not feel they have free choice. Pulled between the desire to work and the desire to be home to raise their children, they find no easy solution in the business world outside the home. For many women, therefore, working at home appears to offer an ideal way to satisfy the need to be a 'good mother' and the need to be a career woman.

Lisa Jacobi

When she was twenty-two, Lisa Jacobi would have been shocked to hear that one day she would quit her job to stay at home and raise her daughters. She had just graduated from college and started out as a management trainee with the East Coast offices of American Telephone and Telegraph. She had every intention of climbing her way up the corporate ladder. Over the next eight years, she did exactly that, eventually being promoted to the position of regional manager.

'I had to meet regularly with managers and employees in our stores. It required me to be on the road a lot. I loved my job. I loved working with people. I always felt like I was on stage.'

During that time, she met and married Stewart, an economist working for the federal government. From the outset they had an explicit understanding that in their marriage they were full partners – both in the workplace and in the home. Neither's career would be subordinated to the other's. Since their incomes were virtually identical it was easy to avoid the trap of assuming that the person with the higher income should exercise more power in the family's decision-making. Although Lisa continued to assume overall responsibility for the house, Stewart took on many of the tasks, so that the burdens were more evenly divided than in most marriages. Given the importance of both their careers and the parity of their domestic relationship, it was assumed that both would continue to work after they had children. 'I always thought it wouldn't be a big deal, that somehow I'd juggle everything', said Lisa.

When she was thirty-two, Lisa gave birth to their first daughter, Ellen. To Lisa's surprise and bewilderment, becoming a parent caused her to re-evaluate their situation. While on maternity leave she began to take a hard look at her job and its travel demands. Despite her hopes of continuing to work she had to accept the fact that 'a baby doesn't adhere to any type of schedule'. Her company offered no childcare assistance, so she spent the first months of her maternity leave scouting for childcare. She was horrified at what she found, finding some unfit 'even for a dog'. A more fundamental re-evaluation was the result of an intense re-experiencing of her own childhood.

'My mom was always there when we got home. We were always able to burst in the door and tell her what we did. I remember how annoyed I was the few times she wasn't at home. "How dare she not be home?" I wanted my mother to be there.'

She began to translate these memories into standards for her own life as a mother and became convinced that staying at home was the only way that she could provide the moral guidance she wanted to give her children.

'If you have a set of values that are real important, you want to be able to transfer them to your children. If you are away from home forty to fifty hours a week, then someone else is influencing your children a lot more than you are.'

She and Stewart came to the conclusion that one of them had to stay home. Stewart was willing to be the prime nurturer and would have been able to, since he had accrued extensive leave time. But to add to Lisa's confusion, she found that she could not accept that possibility or its implications.

'I know that this feeling of mine isn't innate, that it comes from all the conditioning I had as a kid. *Mothers* were always the ones who stayed home and took care of the kids. When Stewart offered to do that, I could not accept the idea that he'd be the mom and I'd be the breadwinner. I just couldn't reconcile that switch within myself.'

But she was not sure she could give up her career and stay at home, either. Therefore, she approached her employer to see if they would let her work at home. But the novelty of the situation – and the perceived difficulties of supervising her from a distance – prompted them to say no.

It became clear that if she wanted to stay at home, she would have to give up her job altogether. The prospect did not please her; in fact, it terrified her. Her career meant so much to her. It had provided her with a sense of power and status, and it had given her the income that she felt was an important factor in ensuring the equality she felt with her husband. The regularity of a salary provided a sustained sense of self-esteem and independence.

'I had had a paycheck since I was sixteen. Now, all of a sudden, I had to decide for the first time in my life not to have a job and to

66

depend on someone else for support. I know that there's no logic to it. I know that when you're at home you work, but I think that having that paycheck in my name was real important. I just knew that I needed to feel like I was contributing to the support of our household. I've always felt like that – even as a kid. I just can't get the words out to ask for money.'

It was a difficult and confusing time for her. She was unsure about giving up her job but could not see how she could maintain her dual and competing allegiances to her employer and her child.

'I didn't think it was fair to go back to the company. I had such loyalty to them. Before I had my children, I was able to commit long hours. After Ellen was born, I couldn't commit the same type of hours. I had to be with her. I knew that I just couldn't give the company 100 per cent like I had before.'

Many career women feel that if they go back to work for their company after the birth of their child they can no longer give as much – in hours or emotional commitment – as they did before. In their minds, anything less than forty to fifty hours a week is unacceptable; an attitude apparently shared by most employers.

Although her employer offered Lisa several jobs during her maternity leave, none would have fitted her schedule as a parent since all required extensive travel and overtime. Although pleased that the company obviously did not want to lose her – she loved working for them and wanted to continue – she saw no alternative but to quit, since they offered no middle ground short of a full-time nine-to-five commitment, which would have made it impossible to give what she needed to give to parenting. Had her employer had the flexibility to offer Lisa job-sharing, part-time work, or telecommuting, they could have retained the skilled services and invaluable enthusiasm of an experienced employee.

Instead, Lisa felt she had to choose between herself and her child. She quit her job.

'I went through what I guess a lot of women go through – wondering if you're doing the right thing. Are you being selfish if you work? Are you giving enough time to your kids?'

Although she left her job, Lisa knew she could not give up the satisfaction, financial and otherwise, of working.

Working at home seemed to offer what she needed. She could earn money and avoid being typecast completely as a homemaker, a role she had assiduously avoided her entire life.

One of the most striking aspects of Lisa's search for suitable home-based work was her loss of confidence and direction. Although she had over ten years of business experience, she initially saw no way to translate that experience into a home-based business. Instead, she began a frantic search of newspaper ads to find any kind of work she could do at home.

> 'I wasn't proud. I'd do anything. I just needed to prove to myself that I could do something at home and use the skills that I had spent all those years acquiring. I found a second-hand store through the newspaper that wanted someone who could call up people and ask for donations. They hired me and I did it two hours in the morning and two hours at night. On Sunday nights, Stewart would help. I only made a little money, but that was all I needed to know – that I could do something at home.'

Lisa used this telephone job to gain enough confidence to figure out her next step. She became convinced she could start her own business. Her husband's personal computer seemed the natural starting point for one, and she knew Stewart would show her how to use it.

The computer has been hailed by some as the basis for a greatly expanded home-based work industry in the future. In Lisa's case the computer was essential and made her business possible. But my research indicates that in the vast majority of cases, the computer plays a minimal role in the initial decision to work at home (Christensen 1986). The first decision to work at home is typically based on family or business reasons. To focus so much attention on the computer is to ignore the real reasons why people are working at home – as Lisa's case illustrates. Her decision was tied to her values regarding child-rearing. Although the computer was an important tool for her, that is all it was – a tool.

Stewart was enthusiastic about Lisa's desire to be at home, but was sceptical about her idea that the computer would be the best cornerstone for a business. He was not impressed when, in the middle of the night, very excited, Lisa shook him awake and blurted out, 'Mailing lists!' He could not believe anyone would pay her to put together lists of names. He remained unconvinced until Lisa secured her old employer as her first client. From that moment on he was her biggest supporter.

Securing her former employer as a first client gave Lisa the confidence she needed to try to develop a direct mail business. She also offered word processing and data entry services to local companies that were too small to buy their own computers. She began her business with one personal computer, but within six months she added two more personal computers and two printers, locating everything in a spare upstairs bedroom that she converted into her office.

Although her new work bore no resemblance to her old job, she threw herself into it with the same devotion. She would work from nine to five, five days a week, at minimum. At first she took care of her daughter herself, but as her business grew she realized she needed help. Fortunately, her parents lived nearby and were willing to assist. Her father had recently retired, so each morning at nine, both her parents would arrive and assume full responsibility for Ellen. Lisa would head upstairs to her office and work until noon, when she would break for lunch, usually coming downstairs with a bag of laundry in hand. 'I'd put it in while we had lunch, and then I'd take it back upstairs after lunch.' She would work in her office until five.

Stewart's experience with computers was important. He would help out at night and on weekends, taking the information Lisa had got from a customer and setting up the necessary computer programs. More often than not, the two of them would work side by side at their computers after Ellen went to bed.

Lisa's family grew as quickly as her business. Within the first full year of business she gave birth to their second daughter, Jessica. However, as a self-employed business owner, she had no maternity leave and no back-up staff to relieve her of business responsibilities. The day she went into labour with Jessie, she got up at 4.00 a.m. to complete a job that had to be done. While she was being wheeled into the delivery room, she was reminding her husband what clients to call to inform them that their work would be delayed. Two days after the birth, she was able to sit at her computer and finish off a project. For all the advantages of self-employment, there is none of the protective coverage that some employees have, like maternity leave. Furthermore, running your own business is time-consuming.

Although Lisa's home-based business provides immediate and sustained access to her children, it has not resulted in radically increasing the blocks of time she can spend with them. In fact, she has a constant struggle maintaining some boundaries.

Having a separate office upstairs has helped Lisa in her efforts to keep her work activities separate from her family life. Nonetheless, the two often blur. Her older daughter often cannot understand why Mommy wants to be left alone.

'Sometimes Ellen stands at the door and pounds. There are times I am in the middle of a telephone conversation, and it's just not appropriate to put the receiver down and growl at a two-year-old. At that time I try to block out what is going on outside and hope the customer doesn't think I am beating someone. Other times I need to go discipline Ellen because she can get out of control. She can work herself up to a fever pitch, and then we've lost the rest of the afternoon. As long as she knows she isn't locked out of my office and she knows she can come and go, she loses interest. The problem comes when she knows the door is locked and she can't come in.'

Lisa has tried to make room for Ellen as much as possible in the office, including setting up a miniature work area for her.

'My little girl has her own little desk and yellow phone over in the corner of our office. She says she has her own company and she named it after me, "Lisa's company". Sometimes I will get on the telephone and she gets on her yellow phone, and she pretends she is talking to her customers. I have one client, Mr Matthews, who is a real pain. So she gets on the phone with Mr Matthews and talks to him all the time.'

Despite some of the problems, Lisa believes that she is providing an important role model for her daughter.

'Ellen is old enough to be getting a clear picture of what work is and when Mommy has to go to the office. Since she has her own little company, she keeps busy putting labels on envelopes. When she and Grandma are doing that, then she has to come upstairs and get more labels. She is very professional and really self-assured. I don't know if that is just her personality, or the fact that she knows I'm here.'

Lisa has tried to keep her operation small and private – and has, for instance, discouraged clients from coming to her home. But as the business has grown larger, new pressures and problems have developed. Periodically she finds it necessary to hire temporary help, which has been difficult. For one thing, she has discovered that for security reasons most temporary agencies will not send their workers, mostly female, out to private homes. When she does manage to get temporary help, it can create chaos. Since her upstairs office is too small for the additional workers, they end up spreading out through the house. For example, one time Lisa was under a tight deadline to send out a mailing. She had three temporary workers, plus her mother, hand-labelling envelopes at the dining-room table. Rubber bands and red dots were all over the floor and the baby had to be kept in the living room with the gate up.

'Jessie started screaming – babies know when you are desperate, that's when they start acting up. Then the phone rang in our upstairs office. I went to get it and Ellen came with me. While I was talking to this customer, she knew my defences were down so she just started slapping labels on the walls and windows. There was enough sun that day to attach them permanently. I still haven't gotten the razor out to take them down.'

Despite such difficult moments, things seem generally to be working out for the family, but for Lisa the situation has not been without costs. She went from a high profile, socially complex role as a corporate middle manager to working alone on a personal computer in an upstairs bedroom. With the exception of her parents, she lost adult companionship. 'There's no one in this neighbourhood under fifty. All the younger people are out working.' This has been particularly difficult, since she used to depend on her workplace for friendships:

'All my friends were the people at work. When I started working at home, I really missed the adult companionship. It was funny, but all I did was talk about my old company for the first nine months after I left. If someone came over, it was just so natural to get into that conversational pattern of checking out who was doing what and who was responsible for what. It's like an addiction.'

Lisa has also lost the luxury of 'going home' and leaving the job behind.

'Sometimes when you work in the same place you live, you get the feeling that you never get away. There is a need to get out and get away to break the monotony of being in one place for such long periods of time.'

In her worst moments, Lisa even longs for the crowded, dirty commuter bus she used to take to work. Those thirty minutes at least gave her some time alone. Now, as soon as she walks out of the office and down the stairs she is home. Even though she runs a business, being home has also made her feel vulnerable to being stereotyped as 'just a homemaker'.

'Because I am a woman, people imagine me wearing an apron. I've heard it in people's voices over the telephone, when I tell them my office is at home. They can't understand that I can have a business and not be making cookies or fudge. They see it as the same kind of thing. They don't see professionalism in it. They might see it when the product is generated, but they don't quite understand. I resent it. I don't get this kind of reaction from women who have had children and may have been torn leaving their children with a baby-sitter. But I do get it from men, even my own brother.'

There is also a paradox that Lisa has recognized, regarding the whole idea of children and home-based work. She quit her job to be home with her children, but she typically works eight to ten hours a day to make the business a success. It turns out that working at home did not add tremendously to the amount of daytime she spends with her daughters. Nonetheless, she feels it provides her with immediate access to them in case of an emergency. She is still there, even when she is working. When she considers the situation she laughs and says she sometimes worries that her children will grow up saying that all their mother ever did was work.

The demands of the business have also created some unanticipated strains in her marriage. For one thing, Stewart and Lisa have very little time together in the evening just to relax.

> 'I quit at five sharp, then the evening routine starts. My parents leave, I start dinner, run around and put in the laundry, and Stewart arrives home at six. We eat and put the kids to bed. Then we both go back in the office at eight thirty. I usually poop out at ten thirty or eleven o'clock, and he usually works a lot later.'

For the most part their personal time together consists of working side-by-side at their computers.

Even though her home-based business has not proved to be ideal in every way, Lisa has decided that it is the best solution for her now. She likes being available to her children, and she prefers to be self-employed. She now says that she will probably never return to the corporate world.

Her home-based business has evolved into an entire lifestyle, one far different than she had ever imagined for herself. She lives and works as part of an extended family – three generations are in her home every day. It seems likely that the business will evolve even further. Stewart is now considering quitting his job with the government and going into partnership with Lisa. If that happens, her home-based work will turn out to have been the cornerstone for a fully-fledged family business.

Older women

The women we have heard from thus far were born in the 1950s and started to work at home while they had young children. They tell only part of the story about women and home-based work. There is another generation of women, born in the 1930s, whose stories are quite different. These are women who started home-based businesses later in life – after their children were grown and gone.

Many women find it difficult to enter or re-enter the work force when they are in their forties and fifties. Many jobs seem trivial or demeaning after decades of responsibility and being in charge of one's self and family. In the past the tremendous energy and enthusiasm of women at this stage in their lives might have been devoted to voluntary work. Now they want to direct it to wage-earning work, and often harbour a dream of what they would like to accomplish. A home-based business offers some an attractive way to achieve such dreams.

Two women's experience with working at home are profiled. Like Anne and Lisa, their decisions about work were intricately tied to circumstances in their marriages and associated with a particular point in their life course.

Janet Tillman

Born and raised in southern California, Janet Tillman dreamed of a life as a fashion designer, but an early marriage cut short those dreams. In 1952, at the age of nineteen, she wed her high school sweetheart, and they moved to Fresno, California, where he worked as a production supervisor at a small manufacturing firm. She worked as a secretary until she became pregnant with their first child in 1954.

'I come from the old school. We women didn't have hard choices. Family always came first with me and I didn't have to find an alternative to being a full-time wife and mother. I was very content to be home.'

She and her family could live the 1950s American dream – a house in the suburbs and two cars – on one income. Middle-class mothers of young children were expected to be at home, rather than outside holding down a job.

Throughout the 1950s and 1960s, Janet led an active and demanding life as the mother of three, feeling powerful and respected in her roles as full-time wife, mother, and homemaker.

'I was in command – I was responsible for my entire family and had to make sure that everything ran smoothly.'

She did all of the cooking and cleaning, was active in her children's schools, and regularly served as a den mother or Brownie leader. Her voluntary work on behalf of her children provided her with opportunities to use and develop her natural organizational and leadership skills. It was a lifestyle that suited her personality and their family circumstances.

As her youngest son reached adolescence in the mid-1970s, Janet faced the fact that her responsibilities as a mother would shortly end.

'I knew I had to start preparing for the day when I would no longer be a full-time housewife and mother. I am a high-energy, take-charge person, and I knew that I had to find something to do after they all left home.'

Rather than expanding her voluntary work, Janet decided to get a paying job, but insisted that it be only part-time.

'I couldn't work eight hours a day with all my responsibilities at home. My energy would not have spread that far.'

To her surprise, this take-charge, confident woman found that the market-place was not ready to greet her.

'Employers would say to me, "You haven't worked in nearly twenty years, so you don't know anything about office procedure." But I knew a lot more than I was given credit for.'

Her frustration grew to indignation that employers did not value the skills required to manage a home and family successfully.

'Job applications should have recognized those skills. If employers ever did an inventory of all of the skills used by housewives and mothers, they'd see an incredible number of demanding skills. Instead, they wanted "marketable skills" and they said that I didn't have them.'

Eventually Janet was hired as a part-time file clerk in a local accountant's office. To her mind, it was better than working as a sales clerk or waitress, but her new job was far from satisfying.

'I found myself in a job where I worked for a man young enough to be my son and that wasn't easy to take. I had all this experience and knowledge and I had to work for someone who knew less than I knew.'

In her home she was a respected, powerful person, but in the office she was a powerless subordinate. This contrast in status was hard for her to take, but she decided to tolerate it as long as she had children at home who required most of her energies.

As her youngest son approached his senior year in high school, Janet decided her time had come.

'One day I sat down and said, "Hey, what do I really want to do?" For the first time I had the privilege of doing what I wanted to do and the chance to be happy in my work. I didn't care what I made so long as I loved the work.'

Her attitude toward a job mystified her husband, Gordon. Having grown up in the Depression and been taught to be the breadwinner, he could not understand placing a higher premium on work enjoyment than on money; his responsibilities had taught him otherwise. For years he had supported a family of five, including during some very tough economic times.

Janet believes that their different attitudes toward work are based on their different family responsibilities.

'Gordon has to support the family so he equates a job with money – making a certain amount of money in a certain amount of time. But I don't. Whatever I make is great, because I don't have the pressure of earning $1,200 a month to pay the bills. If my earnings go up or down, it doesn't matter because I don't have to support the family. He does.'

Although they have had disagreements about her earnings, Gordon and Janet have an openness in their marriage that allows them to air their differences, rather than letting them fester beneath the surface. Although Gordon will never entirely understand her views, he no longer feels he has to. In his mind, Janet has fulfilled her important responsibilities – to raise three healthy, happy children – and is entitled to do something for herself. For Janet that has meant creating a work situation that mirrored her experiences as a homemaker.

She wanted independence to control how and when she worked, and she wanted creativity – both of which she had enjoyed as a full-time wife and homemaker. Her future as an employee looked bleak; she had little interest in working for the next twenty years as a file clerk. Self-employment held more opportunity for her than did outside employment.

A crack typist before she got married, Janet decided to build a secretarial business. 'I figured if I could organize someone else's office and do all the typing, why couldn't I do it for myself?'

Basing her business at home offered her some advantages. Since she would have virtually no overheads and limited requirements for capital investments, other than a typewriter, the home provided the most financially sensible location. But there may have been an additional symbolic reason for starting the business there: her power was in her home. Janet had felt powerless in the business world beyond her front door, and she was much more comfortable establishing her business in her home than outside it.

Janet methodically laid the groundwork for her business, drawing on the organizational and leadership skills she had developed as a home-maker and volunteer. She spent hours in the local library doing research on typing businesses, then turned her attention to her community, surveying local businesses to determine the potential market for her services. She priced her services by calling typists listed in the yellow pages of the telephone book and pretending that she was a potential client who wanted estimates for particular projects.

Having done her basic market research, Janet then attended a one-day Small Business Administration (SBA) seminar that provided information about bookkeeping, record keeping, and taxes for small businesses. The seminar provided the final boost to her confidence. She knew it was time to claim her space at home.

Unlike women who set up their typewriters in the kitchen, Janet was definitive about the need for a separate office, cut off from the traffic of the home.

> 'I came home from the SBA seminar and my youngest son was in his room packing for college. He had boxes all over. I walked in and said, "Well this is my office." He said, "What? Give me a break, I'm not gone yet!" I said, "You're gone, this is my office." I could just envision it.'

The room suited her needs perfectly. It was on the ground floor, off the living room, with separate doors leading into the kitchen and the hall. Its large windows afforded an expansive view of their tree-shaded garden. It offered the privacy and the light that she wanted, and Janet felt no ambivalence or conflict about usurping it for herself. Her son would always be welcome at home, but just as he was ready to start a new phase in his life, so was she.

Establishing the office was the first tangible symbol that she was a business-owner. Within days, she ordered business cards and printed flyers, confirming the fact. She wasted no energy in distributing her flyers to local businesses.

'I took the flyers around and said, "I am starting a home-based secretarial typing service. If you have overflow work, call me instead of a temporary agency." The local merchants took me in like a child and said, "I'll be glad to refer work to you." I began to get walk-in clients from these businesses and students from the local university.'

As the business got going, she enjoyed the money she was earning but disliked the lack of control she had over the pacing of her work. After two months of unreliable piecework, she decided to find a way to get steady contract work.

'I started to think about where I could go to get consistent work. It dawned on me that court reporters need transcribers, so I got out the telephone book and under the yellow pages, it says, "Reporting, Court". So I started calling agencies. An agency five blocks away said, "Yes, we need a transcriber." I didn't know anything about transcribing but I can spell, so I took the work.'

Janet's organization, self-confidence, and ability to push on and adapt were rooted in the fact that she had felt strengthened by her roles as wife and mother, and not undermined by them. The respect and dignity she felt in those roles carried over to her expectations and abilities in her work. When she was dissatisfied she re-evaluated her needs and developed alternative solutions. If the solutions required risks, she thought nothing of taking them.

The fact that Janet and Gordon can talk about their problems has enabled them to respond to their changing needs at different stages in their lives. Gordon is nearing retirement and wants to take off long weekends for fishing. Janet wants to make a success of her business, which often requires her to spend over forty hours a week working. But the fact that she owns her own business and runs it at home gives her the ability to meet her needs and yet be responsive to Gordon's. She will often work four very long days, so she can take off three days with him. They will pack up their camper on Thursday night and take off. She was able to take a month off so they could take an extended camping trip through Alaska, an unlikely possibility had she continued to work as a file clerk for someone else.

Pat Briggs

'I always did what I was supposed to do. I graduated from college. I got married. I married the man I was supposed to marry. I had my children when I was supposed to. I always did what my mother expected me to do. I never disgraced her.'

Pat Briggs excelled at being the dutiful daughter, and all it entailed. A high school beauty, she was elected senior prom queen, went to college to get her teaching degree so that she would always have something to fall back on, married right after her June graduation in 1956, never having taught, and within three years gave birth to Robert and Janice.

During the next fifteen years, Pat was a devoted wife and mother, an attentive daughter, and a successful manager of a law office, having picked up paralegal training along the way. As she approached her fortieth birthday in the autumn of 1974, she began to suffer chronic pain from ulcers, pulled back muscles, and migraine headaches. She had always had headaches, but the combination of ulcers and back pain nearly incapacitated her.

Pat was certain that her problems were related to stress. She broke with her social upbringing enough to undergo a course of psychotherapy, because she began to think that her discontent was related to being the dutiful daughter. 'I realized that I couldn't deal with all of my mother's expectations for me.'

She also realized that her mother's expectations were symptomatic of the broader social climate in which her mother had grown up and in which Pat had been raised.

'When I think back to the 1950s, we really got caught up in a whole bunch of things that never really gave us time to sit back and think about ourselves. Gals were brought up to believe that you went to college, became a teacher or a nurse, got married a year or two out of college, and then had your kids. Today a lot of gals take the time to be themselves before they get married – if they get married. We couldn't do that in our twenties. If we had done something like that, our mothers would have been carting mountains on their backs to atone for us. I was doing what my mother expected of me because that was what her mother had expected of her.'

Through her therapy and discussions with friends of similar ages, Pat grew convinced that the social norms of the times, including her mother's expectations, precluded a woman from making real choices in

77

her life. The lack of choice and the sense that she had no freedom in her life were taking a real toll on her health.

'I finally realized that I was not doing what I wanted to do and I certainly was not doing what was good for me.'

So, at forty, Pat Briggs decided to take the time to explore what it would mean to be herself.

'I decided that the time to be myself and to grow as a person would come toward the end of my life, rather than at the beginning or the middle. But I was determined it was going to be there and that I would get the most out of it.'

Once she began questioning the lack of choice she had exercised in her early life, she became sensitive to the types of choices friends made in their later lives. Her next-door neighbour's experiences proved to be a catalyst to Pat's thinking.

Although a few years older than Pat, Barbara had been brought up with the same expectations. She graduated from nursing school in 1952, got married, had her children, and worked part-time as a surgical nurse until the early 1970s, by which time she was tired – of nursing and of constantly going to school to keep up with the most recent medical developments. When her children reached high school age she decided to try something different and enrolled in a painting class, where she discovered a talent she had never known she had.

She began painting murals on the outside walls of buildings. Her sense of colour, texture, and scale created a demand for her skills, and within several years she was travelling around the country painting modern, expansive wall murals. She loved what she was doing and had never felt so happy and challenged. Her husband was fully supportive of her work, as were her children, by then in college, who were delighted to watch their mother blossom. Her experiences had only been positive and she encouraged Pat to give some thought to what she might really enjoy doing.

Although Pat enjoyed the intellectual challenge of her legal work, she felt it lacked any real creative edge, and she was intrigued by Barbara's experiences.

'I thought it would be really great if I could make my living sewing.'

She had sewn since she was five years old. She was not only fast – able to turn out a dress and sometimes two a week for family and friends – she also had a good eye for fabrics and design. She always designed her own clothes, using a pattern only for basic proportions. Barbara's experiences inspired her to toy with the idea of selling her work.

Once she was open to this possibility, an opportunity presented itself. While in the Caribbean on holiday, the manager of a small boutique in which she was shopping asked to buy the coatdress she was wearing. When she realized that Pat not only had sewn it but had also designed it, she offered to buy as many as Pat could produce. That experience gave Pat the first taste of making money from doing something she loved. But, like Janet, she was slow and cautious. She decided to start sewing enough to sell at craft fairs, then see what the response would be, before making any deeper commitment to a custom sewing business.

She kept her job at the law office, working thirty hours a week, but spent an increasing number of hours at home sewing, often working a total of fifty hours or more a week. She loved the satisfaction of working on something that stirred her soul.

'I probably have some repressed urge to create, and sewing is creative. I can get up at four in the morning if I want to do something that just suddenly comes to my mind. I am not restricted. I have no boundaries. I am not filling some niche that someone else has set for me. I am doing what I want to do at my own pace in my own time.'

In the process, however, Pat began to spend less and less time on household matters such as cooking and cleaning. She began to do some travelling to craft shows. When her husband came home at night there was not always a hot meal waiting for him; in fact, sometimes there was a note telling him what was in the freezer. If he made social plans for them that conflicted with her work schedule she would speak up, insisting he go but that she had to stay at home. If he was upset, she was not as likely to make the effort to cajole him into a good mood as she had before. In order to keep her job and pursue her sewing, Pat had changed in ways that did not please her husband. He was not ready to shift in his expectations for their marriage.

'Throughout my entire marriage, I had subordinated to him. By starting something on my own, by having to travel to market my work, and by being successful, I changed. I was no longer what he had in mind for a wife. I was no longer able to come home after work and wait on him. I was becoming very different from the wife he had known for twenty years. I was more outspoken, more independent, and less submissive.'

Although she liked those changes in herself, her husband did not. Barbara's husband had encouraged and supported her evolution, but Pat's husband was threatened, and their marriage broke up. Pat received no child support or alimony, since her children were already in their early twenties and she had worked for years in the law office. This change in her financial situation forced her to give serious thought as to how she would survive, particularly looking ahead to her old age.

'I became more calculating about sewing as a business, because I have no pension plan at the office. In my particular administrative category there are no fringe benefits.'

Like so many women, this situation had never bothered Pat when she had been married, because she had been covered by her husband's health plan and would have had access to his pension funds later in life. She had never anticipated a divorce and having to take fringe benefits into account.

'But after my divorce, I suddenly looked at myself and I thought, "Where am I going to be when I reach sixty-five and there is nothing but this virtually dormant Social Security system?" That's when I decided to build my sewing into a big business. I felt that it was something I could build on as long as I started then and gave it enough time to grow.'

Pat knew that she would have to be patient about the time it would take to develop her sewing into a fully-fledged business, anticipating that for the next six years she would keep her job at the law firm and build the sewing business on the side. She needed the law job for its stable income, and the sewing and design for its creativity.

Although she works long hours and bears the financial responsibility for her daughter's college education and all of her own monthly expenses, she has an energy that is generated by living life on her own terms. Unlike her early years, where she fitted her life into the niches created by her mother's and husband's expectations, Pat now carves out the contours according to her creative needs.

Although any divorce is difficult, Pat maintains a friendly relationship with her husband, feeling that they unfortunately just outgrew each other. Interestingly enough, she has grown into deeper and more satisfying relationships with her son and daughter. The stronger she has become in her own right, the stronger her relationships with her children have become.

'My children tell me that I am an entirely different person, that I am more confident and that I amaze them with what I am doing and want to do. I love it. Until now I always felt that there was such a huge gap between us – I was the mother and I was stuck in this little hole. Now I realize that it was because I wasn't happy with myself. Now I am. I have my own interests. I am creating something on my own.'

Despite the fact that she loves what she now does, she is not convinced that as a young mother she could have undertaken the kind of commitment that her sewing business requires.

'As a mother you are very much aware of your responsibility for your children. When you have the responsibility for someone else, whether a child or an ageing parent, you can't really take any risks, because if you take a risk and you fail, that risk and failure will affect them. And you can't do that; you have too many people depending on you.'

Part of Pat's belief about the constraints of motherhood may also be tied to the fears that she herself experienced as a mother of young children. She was the type of person who had great difficulty in relating to small children.

'I would have been marvelous with my children if they would have [sic] been born as adults. I was so nervous. I was so afraid that I was going to do the wrong thing. I was so afraid that I was going to put them in some life-threatening situation or do something awful so that they'd turn into idiots by the time they were thirty-five. I never had the time to enjoy them.'

As she grew confident in herself, as her children got older, and as she assumed total financial responsibility for herself, Pat began to see ways to chart new terrain for herself.

Pat no longer has ulcers, headaches, or backaches. If she is tired, it is because she works long hours. But she loves the life she has created for herself and calls her home, where she does her sewing, the 'Yankee Clipper' – a fast-sailing American ship. She sees herself as the clipper's captain. 'I never dreamed that I would live life on my terms and be doing something creative. But I am much happier.'

Women who gave birth to their children in the 1950s are as different from one another as women who gave birth in the 1970s. Some, like Janet, wanted to stay at home and were happy and at peace with the decision. Others like Pat were not content.

As the women got older, however, each wanted more satisfaction from paid labour than they were able to get in the conventional workplace. Each woman turned to a home-based business as the best way to achieve the independence, creativity, and power that was unattainable for her in the conventional world of work. Both had worked too hard in their lives, raising their children, to sell themselves short in later years. For the first time in their adult lives they have found the energy to devote to a work effort that pleases them, and they want the satisfaction of that experience.

These two women are survivors and probably will always be happiest working on their own. But for other older women, who would like careers in more conventional workplaces and professions, it would be good to examine how the business world's structure of advancement and success is tailored to the lives of men – particularly men who do not have childcare responsibilities. What the stories of these two women indicate, perhaps more powerfully than anything else, is that there should be another way to chart a career, which would take into account the periods of child-rearing and utilize the skills and energy of those who raise our children. Unlike men, who typically are ready to slow down in their work as they reach their fifties, these women are ready to hit their stride.

CONSEQUENCES OF WORKING AT HOME

To understand the consequences of working at home, we must address two factors: how the micro-scale geography of the home can accommodate paid labour; and how the employment status of the home-based worker affects her health and pension coverage.

Micro-scale geography of the home

Most homes in the United States are not designed to accommodate wage-earning work. As a result, when a woman brings paid labour into the home, she must establish space and rituals appropriate to the demands of the work and the conditions of the family.

Information-based work usually requires separate office space. This is particularly important when a woman starts her own business, which typically involves clients or temporary workers coming into her home. The design dilemma most often mentioned by the women I interviewed was how to carve out separate office space that affords access to clients without allowing them to invade the more private areas of the home, such as the bedrooms. According to the women, the ideal solution to this dilemma would be a home office that has a separate door to the outside, so that clients or workers do not have to come into the home at all. This ideal is rarely accomplished. Instead, the bedroom is the most typical room chosen for an office. How well that worked in the cases we looked at depended on the location of the office within the home.

In Janet's case, she was able to use a ground-floor bedroom that was located directly off the living room. When necessary, therefore, a client could be brought through the front door, into the living room, and then directly into her office, without going into other areas of the home. Lisa, on the other hand, had no ground-floor space available and had, therefore, to locate her office in a spare upstairs bedroom. Although this afforded some benefits from a family perspective which will be discussed later, it was not perfect from a work perspective. She did not want clients coming into the bedroom area of her home, so she discouraged them from coming into her home. But she did occasionally have to rely on temporary help. When that occurred the living room and dining room, which usually were used only by her children and their grandparents, were drafted into use as offices, creating a conflict of use between her family and the temporary workers. Other kinds of

82

home-based work such as childcare or home daycare centres generally require more substantial modifications to provide safe and stimulating environments for children.

When a woman works on her own as Anne does as an independent contractor, and as Pat does as a seamstress, she can more easily create her office in a space shared with other family activities. Anne types in the kitchen, while Pat sews in the living room. The problem that can arise is that the setting can present conflicting messages. This is clearly the case with Anne who finds that the kitchen 'talks' to her, demanding certain activities, such as cleaning the refrigerator or preparing the dinner, that she could have ignored more easily if she worked in a space devoted solely to her work.

While a home-based office must be situated to accommodate the needs of the work as indicated above, it must also be located in such a way that the demands of the family can be kept to a minimum. Obviously, this is more of a concern in families in which there are young children. Women in these families need an office space that can be effectively shut off from the sounds, demands, and activities of the children, when needed. While Lisa's office is not situated in an ideal location from the vantage point of work, it is well-located from the perspective of the family. She is able to go upstairs, and when necessary shut the door, so she can work. In addition, since she is off the beaten path of her children's play areas downstairs, she is periodically forgotten by her daughters so she can concentrate and work in isolation. When her older pre-school daughter remembers where she is, she demands access. To create a more permeable boundary between her work and family, Lisa has designed a small office inside her office for her own daughter's use. This way her daughter can come in, but only to work at her miniature desk with the broken calculator designated as her personal computer.

In my larger research project, I have found other ways that women cope with the separation anxieties that toddlers and pre-schoolers experience in leaving their mothers. Some are more elaborate than others. For example, a New Jersey mother of a pre-schooler built separate outdoor steps to her second-floor home office. As part of her daily ritual, she kisses her daughter goodbye, goes outside, gets in the car, drives down the block, only to circle back and sneak into her own office. She has found that her daughter has too difficult a time, at this age, in having her mother at home but inaccessible to her.

What is clear is that when children are infants to pre-schoolers, it is important to have space that can effectively be separated from the home, as well as having support in caring for the children. In my 1985 survey, two-thirds of the professional mothers of pre-schoolers used some form of childcare help for which they typically paid. Only one-third of the women doing clerical work had help, typically on an unpaid basis, such as from a family member or neighbour (Christensen 1986, 1988). When they did not have help, as was the case with Anne when she first started working at home, they would work after the children had gone to sleep. This type of regimen proved exhausting and most women do not continue this over any long period of time.

By the time a child is older, the mother can more easily work at home, either because the child can play without constant supervision or involvement, or because the child is away at school. As a result, the conflict between work and family does not need to be regularly played out in the landscape of the home, because the child is removed from there during the times the mother is most likely to have to work.

Clearly, the women who encounter the least conflict between work and family are those whose children have grown and moved out, or those who do not have children. Pat was able to set her sewing machine up in the living room and work without conflict because she lives alone most of the time. She shares her space with her daughter only when she returns from college for vacation. Janet was able to claim space for herself only when her son was on his way to college.

As women move through the course of their lives their ability to work at home without conflict from the demands of the family changes. Clearly, the time of maximum conflict is when children are young. The irony is that this is the time period during which women are most apt to report that they want to work at home. Older women without children typically fall into working at home when the workplace fails to respond to their needs.

Health and pension consequences of working at home

The health and pension consequences of working at home are tied to the employment status of the workers.

Healthcare

The major healthcare penalty that a home-based worker incurs has to do with the fact that she is most likely a self-employed worker. In the United States, such workers are not covered for health insurance in the same way that employees are. They must provide the coverage for themselves. My 1985 survey results indicate that home-based working women are most likely to receive health insurance coverage through their husbands' plans. In the event of their not being covered that way, then they are most likely not to be covered at all. Single or divorced women working at home face a high probability of not being adequately covered for health insurance.

Older women who work at home and assume that they will be

adequately covered by Medicare will find such care lacking, at least according to national statistics. Nearly two-thirds of Medicare recipients carry supplemental insurance. Some buy it through groups, others take out individual policies. As a result, they pay more and may experience more limited options in the coverage that they can get. In other words, a major cost of working at home relates to the additional costs of healthcare, or the emotional, as well as financial price, one must pay if not adequately covered for a health crisis.

Pensions

In general, pension coverage for older American women is abysmal. According to the Census Bureau in 1986, only one out of every five women over the age of sixty-five received an income from a pension, whether that pension was private or public, or collected from her own or her spouse's coverage. The picture is slightly better for the younger cohort of women. According to the Current Population Survey of all employed women in 1983, only three out of eight received pension coverage. In addition, women's earnings from pensions are substantially lower than men's.

Women who work at home are at even more risk of inadequate pension coverage than are working women in general. While the majority of working women are employed and, therefore, have some likelihood of getting an occupational pension, the majority of home-based working women are self-employed and unless they establish their own retirement fund, they will receive only Social Security income after they reach sixty-five.

Since women who work at home do so predominantly on a self-employed basis, they are not automatically covered by a pension plan. It becomes incumbent on younger women to develop some independent retirement plan as a way to protect themselves in their older age – otherwise these women may find themselves in a position comparable to Pat's, that will require them to work well past retirement age. Although Pat has created work that is intrinsically satisfying, others may not and they will be in the unfortunate situation of having to work longer on less than pleasing work.

POLICY DIRECTIONS

A number of public and private initiatives might be pursued to ensure that working at home proves more advantageous for working women.

Federal directions

Despite some recent calls by organized labour, I do not believe that the US federal government should prohibit white-collar work from being done at home. Over half a million women are currently gainfully employed at home, and although the arrangement may not be perfect, evidence does not warrant that the arrangements be outlawed. The federal government could and should provide support that better protects those women who work at home, or at least should enhance their ability to make it a profitable arrangement. First, they could provide more funding to the Small Business Administration (SBA) to provide outreach and training in the development and running of small businesses from the home. Second, they could better protect women in terms of pension coverage by encouraging portable pensions or shorter periods for vesting. As noted earlier, many of the financial penalties of working at home for older women have more to do with their employment status and their related opportunities for health and pension coverage than with working at home *per se*.

Private directions

The private sector could go a long way to ensure that women at different stages in their lives have more choice in the labour market than is currently available. Such an increase in choice, in effect, would help to ensure that the woman who then decides to work at home is doing so because she really wants to and not because she has no other alternative. Some options for women that the private sector could pursue more aggressively include the following: part-time work congruent with career experience; telecommuting, which allows employees the opportunity to work at home; job-sharing, that allows one job, on a career path, to be shared by two people; phased retirement, which enables older workers to phase out of full-time and into a more part-time arrangement as they reach sixty-five years of age; and temporary pools internal to corporations whereby younger or older workers can re-enter the workforce when income or other reasons necessitate. These same options should be extended to men as well as women, so that the entire household would have more flexible choices in the workplace.

Conclusion

Any discussion of women and home-based work has to take place within the prevailing options these women face in today's labour market. These options, all too unfortunately, are not great.

Many of the solutions that will make working at home better for women are, in fact, likely to be options that would make working anywhere better: more choice regarding flexible schedules; better and more portable pensions; and more adequate healthcare coverage.

4

GROWING GIRLS/CLOSING CIRCLES

Limits on the spaces of knowing in rural Sudan and US cities

Cindi Katz

Social power is reflected in and exercised through the production and control of space. These sociospatial relations are gendered and vary across the life course, riddled by differences associated with class, ethnicity, race and nationality. From 'dad's chair' to occupied national territories, the spatial forms of control are charged with and interpenetrated by political-economic power, cultural meaning and personal significance. These conjunctures are neither stable over time, nor distributed evenly across space. This chapter explores their form and significance at particular periods and transitions in the life course of females, emphasizing the shifts from childhood to youth and womanhood in two divergent settings – rural Sudan and urban United States.

The notions that access to and control of space are greater for males than females and increase with age, at least until middle adulthood, are so commonplace that they remain largely unexamined. Their significance in everyday life has been under-thematized and thus little explored in social science research. While Western social scientists have been vocal about restrictions on women's mobility in Islamic cultures – explicating and condemning purdah as a monolithic restraint on women – the constraints on women's access to space in the West have been less well recognized. But in many ways these are as formidable and as systemic as those associated with Islam. The spatial forms in which social control is manifested are emblematic of the contradictory intersections between public and private spheres in everyday life. This chapter addresses some of these issues cross-culturally to make more vivid the forms, variations, meanings and interpretations of different conjunctures of space and power over the life course of females.

Evidence from my research among children in rural Sudan contradicts the propositions that boys exercise greater control over space than girls; and that access to space expands with age. First, the home range (the distance children travel away from home unaccompanied by adults) of boys did not appear to be greater than that of girls during early and middle childhood. Second, while boys' range continued to increase as they came of age, girls' access to space became constricted at puberty, and the spaces females controlled were close to home, at least until they passed their prime child-bearing years. While in the United States it is generally the case that spatial range increases with age for both males and females, the reasons undergirding it and the differential consequences it engenders for men and women across the life course bear greater scrutiny. Fear of personal injury or violation may be the purdah of the industrialized West, effectively curtailing girls' and women's access to public space. Yet Western scholars continue to train their focus on purdah and its restrictive implications for women's lived experience, ignoring how girls and women in the industrialized West are similarly constrained in their access to and control of space.

In examining these relationships across the life course, both within each setting and between them, this chapter explores the diversity of females' spatial experience, addressing some of the enduring consequences of women's restricted access to and control of the environment. The chapter contributes to breaking down stereotypical notions of purdah by demonstrating some of the asymmetries of purdah, and similarities between it and other strategies by which women's movements are restricted.[1] In drawing these comparisons my intent is to deepen our understanding of how gendered and hierarchized power relations are manifested in the spatial forms and practices of girls' and women's everyday lives, and to reflect on their significance, both in women's experience and for the social formations within which they live their lives (cf. Monk 1992).

CHILDREN'S SPATIAL RANGE IN RURAL SUDAN

In Howa, a rural Sudanese village of about 335 households where I conducted research in 1981, children of both sexes roamed widely and freely both within the village and to fields, pastures and wooded areas at its outskirts 5 km away. The twinned plagues of supervision and lack of autonomy that are increasingly the lot of young children in urban areas of the United States and elsewhere were not theirs. Not only did these children play in and around the village independently beginning in early childhood, but by nine or ten years of age they had responsibility for tasks that routinely led them out of the confines of the settlement. With this responsibility came a fair degree of autonomy in their control of time and space. The tasks for girls included the routine collection of wood and the occasional gathering of food and other resources from areas up to an hour's walk away. Boys also left the village to collect woodfuel and other woodland resources as well as in the course of herding, which was the responsibility of about a third of all boys in Howa at some time during their late childhood or adolescence. In these tasks the children participated in decisions concerning the terrain they worked, often choosing it themselves; were responsible for selecting the paths to and through it; and independently determined when to return to the village each day. Children of both sexes also left the settlement to assist their elders in the outlying farms that ringed the village from 1 to 5 km away. In agriculture their autonomy was more curtailed. Time was controlled by the elders, who set the pace, scheduling and daily limits for each task. The space of work was clearly demarcated within particular farm tenancies.

My research indicated that boys and girls alike shared a rich geographical knowledge that included an understanding of local spatial relationships and an extensive knowledge of the local environment and its resources (cf. Katz 1989; 1991a). In large measure these were gained in the course of their work and play in and around the village. While the content of the girls' and boys' activities differed and was reflected in their knowledge, the arenas in which these activities took place did not. Unlike their American counterparts, these children were at liberty to explore their neighbourhood and its surrounds (cf. Tindal 1971; Hart 1979; Bartlett 1991). Their geographic knowledge was but one reflection of this freedom which children in the industrialized West increasingly lack.

Ironically, while the socio-economics of the contemporary US city increasingly restrict children's movements, the broad socio-economic changes imposed upon Howa by incorporation in the large-scale Suki Agricultural Development Project in 1971 brought about forms of environmental degradation that increased the distance children travelled to provide the means of existence for their families (Katz 1991a). While socio-economic and cultural-ecologic changes in Sudan increased children's home range, albeit in a problematic way, the changes affecting many US cities, such as de-industrialization and local disinvestment, have led to infrastructural deterioration and declines in services that constrict children's access to their environment (cf. Hart 1986; Katz 1991b. White 1990 addresses the same issue in urban Australia).

The implications of these shifts in Sudan are discussed elsewhere (e.g. Katz 1986, 1991a). What I wish to emphasize here are the links between children's geographical experience and both its meaning at other stages in the life course and its relationship to the activities of siblings, parents and other elders. In order to do this, I will first trace out the activities and geographic range of children at different ages, situating these in relation to the work of the village in general.

The settled area of Howa was between 1 and 2 square km. The 335 or so houses in the village were in eighteen loose clusters called *feriq*.

Plate 4.1 Children's life in the *hosh*. In the foreground
girls chop okra (a staple food source throughout the year)
for sun-drying. Behind, a girl grinds sorghum while her
brother and some friends play

These *feriq*, which are analogous to neighbourhoods, consisted of
patrilocal settlements of extended families who shared a semi-enclosed
and socially demarcated space known as a *hosh*, or houseyard, within
which each nuclear family had its own one-room house and separate
grass-enclosed kitchen. From the time they were able to walk, children
moved freely within the *hosh* of their extended family. While most often
they were in the company of older siblings and cousins, it was not

91

Plate 4.2 A woman braids her granddaughter's hair. The child's fifteen-year-old sister (nine months pregnant) sits with them and her seven-year-old brother plays nearby. They are sitting in a newly built brick enclosure within their houseyard

unusual for a one- or two-year-old to toddle across the yard to visit grandparents and other relatives. Given that most daily activities took place outdoors and were focused on the communal space of the yard rather than the public spaces of the street, children were observed easily from around the *hosh*, and thus wandered securely.

From the age of three or four, children delivered messages or carried food and other small items from one house to another within the compound. When just slightly older, they began to run errands around the village – picking up something at the store, fetching small containers of water from the well or conveying messages to neighbours. The age at which these activities commenced depended in part on the children's birth-order position. First-born children tended to assume responsibilities at an earlier age than those with older siblings to do the work. Moreover, while certain chores were passed down, there was a tendency for older children to continue with many of them, sparing their younger brothers and sisters for other activities, among these, playing and school attendance. By seven or eight years of age, all children in Howa did these tasks at least occasionally. At this time children also began to work outside the confines of the settlement, accompanying their elders to assist in the fields, or going with other children to collect wood and other plant resources. By age nine or ten, boys started to herd their families' flocks. Through adolescence, boys continued to work at these and other activities, gradually assuming greater responsibility. But girls' access to the environment outside of the *hosh*, and especially outside of the village, narrowed considerably as they reached puberty and married. While pubescent girls, particularly those without siblings to take on the work, continued to fetch water and wood, they began to restrict their movements around the village – no longer playing in the streets, and rarely running errands, for example. At this time girls began to wear flimsy shawls over their heads whenever they left the hosh. (As with adults, covering themselves, even tokenistically, was a sign of modesty that facilitated the girls' movements through public spaces.) Girls in Howa began to marry and bear children by their early teens, during which time their withdrawal from public space was intensified.

There are historically and geographically contingent interrelationships between men, women and children in carrying out the work of household reproduction. While children throughout Africa and other parts of the Third World are often essential to the provision of subsistence (e.g. Rodgers and Standing 1981; Nag, White and Peet 1978; Cain 1977; Ruddle and Chesterfield 1977), such work is generally the province of women. In places where women rarely participate in tasks that take them outside of their family compounds, the importance of children's work is magnified. Thus, as Enid Schildkrout (1981) demonstrates for urban Kano, Nigeria, where children supported women's trading activities by moving publicly between traders and customers, in Howa children undergirded and in many ways enabled the practice of purdah because they procured the means of existence and accomplished many of the tasks associated with household reproduction (cf. Longhurst 1982 for an example from Muslim Hausa agriculturalists).

In Howa, as elsewhere, women were secluded to different extents, depending upon their class position, life course stage and household circumstances. The work of reproduction, which includes the production, provision and preparation of the means of existence; the production, sustenance and socialization of children; and the production and exchange of social knowledge, is tied inextricably to the work of production and the social relations of production and reproduction that underlie it. While these practices of social reproduction are always in flux – for example in the nineteenth century the spread of public education in the United States expanded the role of schools in the socialization of children, and in the second half of the twentieth century television assumed an important role in the same process – they tend to be stable: the nuclear family remains the primary source of children's socialization in most Western cultures. When these practices are not

93

accomplished successfully, whether by paid labour, unpaid labour or under the aegis of the state, social formations are not sustained. Webs of social power relations are implicated in the particular constellations by which the work of production and reproduction are carried out at historically and geographically specific junctures. Within households, for example, there is a particular balance between men and women in carrying out this work. This balance, as well as the social contract that supports it, differs between and among classes, ethnic groups, nations and individuals. These shift over time, propelled and propelling a characteristic balance between public, private and personal means of ensuring production and reproduction.

In rural Sudan, there were life course, gender and class parameters to these relationships. Most domestic water, and a substantial portion of the woodfuel, for example, were provided not by women but by children and teenagers. When women passed their prime child-bearing years, they also procured water and wood. Depending upon the availability of other sources of labour – including children, men, members of the extended family or paid water-bearers – nubile women occasionally fetched water as well, although those that did tended to be from households of lower socio-economic status. The division of labour between men and women also supported the maintenance of purdah in Howa. Men predominated in agriculture, except in harvest activities; additionally, along with teenage boys, they were responsible for most of the tasks associated with animal husbandry, except milking which was done by women within household compounds; and they cut wood or produced charcoal from woodlots outside of the settled area (cf. Mies 1982 who discusses similar gender divisions of labour and their articulation with capitalist development and the uneven practice of purdah in Narsapur, India). In these ways and others, the work geographies of male and female family members at different phases of the life course were fundamentally intertwined. Just as the work done outside the *hosh* by men, youngsters and older family members enabled the maintenance of purdah for younger women in an extended family, the work women did within the family compound, including food preparation, laundry and child-minding, facilitated and was sustained by the outside work of their relatives.

CHILDREN'S GEOGRAPHIES IN AN ALTERED SOCIAL SPACE

These spatial constructions were undergoing a profound alteration in 1981 as a result of the changes brought about by inclusion in the agricultural development project ten years earlier. Given how tightly the patterns of each gender, age and occupational group were articulated with one another, any alterations within the constellation had individual and collective reverberations which had serious implications for present and future. Of particular significance here because of their gender and life course implications are:

1 the working of women in their households' agricultural tenancies and as piece-workers during the three-month harvest period, given the intensive labour demands of the cotton harvest and the growing local need for cash;
2 the expansion of children's labour time and space, due to the expansion of the cash economy, the degradation of the local environment and their mothers' changing work patterns;
3 the beginning of male outmigration, due to limitations on access to land and other productive resources resulting from the fixed number of farm tenancies allocated to the village.

Each of these spatio-temporal shifts raises issues in which production, reproduction and knowledge intersect. These intersections are associated with particular socio-economic interrelationships between males and females of different ages both with meaning for the contemporary work landscape and intergenerational repercussions.

As the capitalist relations of production associated with the agricultural development project came to predominate in Howa, the growing need for cash led increasing numbers of women to pick cotton and harvest groundnuts both in their families' tenancies and as piece-workers. Here again, the differential social impact of the agricultural project was significant – women from families that were relatively well off worked neither in their own fields nor those of others, and women in the younger child-bearing years almost never worked the fields, no matter how dire their household circumstances. However, by 1981, ten years after the Project was established, I calculated that over 42 per cent of the village households had earnings below subsistence level (Katz 1986: 44). In the face of growing impoverishment the strictures of purdah began to break down. In my village-wide household survey, 29 per cent of women in Howa reported working for cash during the harvest, and this is a conservative estimate. Of these women, virtually all who were from tenant households worked the harvest in their own tenancies as well. It seemed likely that as the commodification and socio-economic differentiation associated with the Project continued,[2] the number of women working the fields – both their own and for cash – would grow. To accommodate this new socio-economic situation spatiotemporally, many families in Howa set up 'camps' and moved their entire household to the fields during the harvest. In this way the tasks of production and reproduction were brought together in a different setting during the

95

harvest, enabling the maximum labour input from all family members, including women with young children.

Children's work also seemed to be intensifying as a result of the twin processes of environmental degradation and commodification associated with Howa's inclusion in the agricultural project. In brief, to establish the Suki Project approximately 2,500 acres were cleared for farm tenancies in the immediate vicinity of the village. In the process, many pasture and wooded areas were destroyed. The Suki Project as a whole transformed 85,000 acres in the area from mixed land use to intensive dual-crop agriculture, putting immense pressure on the remaining wood and grasslands, which had become severely degraded by 1981. This destruction of the local environment required children and others to go further afield and to spend more time to procure adequate household fuel supplies and graze animals. These changes in the local ecology increased the work of both boys and girls. While girls were more prominent in providing woodfuel, children of both sexes made more frequent forays and spent more time collecting or cutting in response to the decline in woodlands and the deterioration in wood quality. The changes in grazing lands brought about by the Project did not so much increase the work time of shepherds, as alter its nature. These changes, which worked in contradictory ways, affected boys who alone were herders. The deterioration in pastures and deforestation in the vicinity increased herdboys' work, especially during the dry season when enormous effort and time was required to find adequate forage and pasturage for the animals. During the wet season when the fields were planted, the work of herdboys may have been lessened as a result of the Project because of the local proclivity to graze animals on the cotton crop,[3] and because fortuitous breaks in the irrigation ditches led to the growth of miniature meadows. Because shepherds' work is characteristically lighter in the wet season, these savings in labour time were insignificant compared with the increases engendered during the dry season by local environmental degradation.

Children also worked more in response to Howa's heightened incorporation into the cash economy and the attendant commodification of many goods and services. Both boys and girls earned cash for their households by selling water, wood and food, for example; and working as paid field-hands. In Sudan's Gezira Scheme, Galal el Din (1977) found that almost 22 per cent of the total labour force picking cotton was under fifteen years old, and over 8 per cent of the paid cotton pickers were under ten years of age.

In many households – again particularly the poorer ones – these changes interfered with school enrolment and attendance. My research suggested that it was the poorest and richest households that appeared to require the most work of their children. In the impoverished households, male and female children's labour was necessary to maintain basic household subsistence and provide cash on occasion. In the wealthiest households the differentiation of household production activities required increased labour of children as well. In these settings there was a process of specialization which affected boys in particular – certain sons worked the fields, others were sent to school, others learned about raising livestock and still others learned about trading. Children contributed in each of these arenas, and through the process were also groomed for their future roles. Under these conditions I did not discern any systematic variation in girls' roles, and the extremely low level of female school enrolment in Howa (only 4 per cent of the girls between seven and twelve years old were enrolled in the village school), precludes systematic analysis of whose daughters went to school.

The third sociospatial transformation of particular relevance here was that by 1981 young men were beginning to leave Howa for nearby towns in search of work. Only 250 tenancies were allocated to Howa when the Project was established in 1971. From the outset the allocation fell short of the number of households in the village, but as new households were formed and tenants remained active, the chance to acquire a farm tenancy grew progressively slimmer for each prospective tenant. As a rule, agricultural projects, like other development efforts, do not take demographic dynamics into account. Nor do planners address the intergenerational consequences of 'development'. In Howa by 1981 this short-sightedness had begun to result in the emigration of young men. This situation was enmeshed with the nature and pacing of life course transitions there.

Tenancies were generally allocated to male household heads, one to a household. New households were formed upon marriage and established more formally upon the birth of the second child. Young men generally married in their late teens, young women in their early teens. As they established their new households, it was expected that they became self-supporting. Prior to the Suki Project, when subsistence dryland cultivation was the practice in Howa, young men coming of age would have been given a small plot from their families' holdings or would have cleared new land at the margins of the cultivated area. As household needs grew, they would extend the area of cultivation into new areas or the fallowed land of their extended family.[4] With the Suki Project, new conditions prevailed. Given the fixed number of tenancies allocated to the village, the fact that the average family in Howa had five children and the early age of marriage and child-bearing (which meant, for example, that many twenty-year-old men had parents still in their thirties and thus a long time from retirement), few young men would have access to land as they began to form their own households.

In the face of these conditions, lured by the 'better' life they witnessed in the towns, young men in Howa were beginning to seek employment in the nearby project headquarters and market towns. Male outmigration often leads to the increased participation of women and children in agricultural production. If this results in Howa, the practice of purdah may be further eroded there. If it leads to more general rural depopulation, however, and households are reformulated in regional centres, male outmigration – which was just beginning during my field study – may ultimately reinforce purdah, because female seclusion is more easily facilitated by town life where it is customary for male household members to do all the marketing, and where, particularly among the middle class, there is heightened social pressure to conform with its strictures. If young couples move to towns without their extended families, as was likely to be the case, purdah may well come to mean greater seclusion than in rural settings where the *hosh* provides a setting that is anything but secluded.

These shifts in women's, children's and men's activities were likely to lead to new sociospatial patterns, not only for men and women over their life courses, but for the village and region as a whole. For example, as women began to work away from the household for certain tasks, the construction of purdah was altered to support the expansion of women's spatial domains. To the extent that women take on tasks previously performed by children – for example, woodfuel collection when the only available sources become too distant – children's time is released and a growing number might attend school. Whether children will work more or less as a result of the shifts underway in Sudan is unclear. Increases and decreases in children's labour are part of a matrix of sociospatial work relations. These shifts were taking place in a political-economic field in which it was becoming increasingly clear to many parents that the new regime of accumulation in Sudan required a new kind of preparation of their children. As they witnessed the quest for work by the first generation to come of age after the establishment of the Project, their best hopes were that government schools would prepare their children for the changing circumstances they were facing.

The population was not passive in the face of external change. In 1984 the village council of Howa sponsored the construction of a girls' school using self-help funds generated through the village sugar cooperative.[5] Cognizant of the complex space–time considerations of household work, however, the village council first sponsored the construction of standpipes throughout the village with the option for individual clans to bring piped water into their compounds. This local initiative, supported by the State, reduced children's and particularly girls' labour time substantially. In many cases it made the difference between attending school or not.[6]

It is not trite to note that school attendance would expand the horizons of girls in Howa. Some of them might go on to high school and perhaps careers as teachers, health practitioners, midwives and social workers. Under present political-economic and sociocultural circumstances, however, most of them would not. More likely, the increasing rate of male outmigration would have a more substantial effect on female horizons, resulting in less metaphoric expansions. In the face of a rising male exodus from the village, agricultural production was likely to be taken over by women in Howa. Under these conditions, the work and learning experiences of young girls in the fields, forests and open lands around the village, would be useful to them in their adulthoods, and their social activity space upon reaching puberty might not be as constricted.

Rural productivity under these shifting labour and sociospatial conditions is precarious. Under the conditions that obtained in 1981, for example, girls withdrew from active participation in agricultural work as they reached puberty and did not learn the overall coordination and sequencing of agricultural tasks as did the boys. Nor did they have the extensive hands-on experience with the full range of farming tasks of most boys in Howa. If agricultural production were to be left in their hands, as they reached adulthood and found their husbands and brothers migrating from the area in search of work, they might not be fully prepared, and productivity might suffer. Conversely, what was equally striking during my field research was how much they did know, how far they had explored and how familiar they were with the local environment as a resource base – and how this knowledge languished, untapped in their conventional work roles and shrunken sociospatial domains. This dormant knowledge was available to be tapped in the new constellations of work and social relations.

SPACE AND LEARNING

Children's knowledge in Howa was rooted and developed in their extensive environmental experiences. Developmental psychologists working cross-culturally have demonstrated a positive relationship between 'self-managed sequences' of work and play and the development of analytic ability (cf. Nerlove *et al.* 1974). Going out alone in the community is an important arena of self-managed behaviour for children. Earlier research by Nerlove and her colleagues revealed direct links between children's autonomous environmental experience and the development of analytic ability, including the acquisition of large 'cognitive maps' and the spatial ability that goes along with their construction and negotiation (cf. Munroe and Munroe 1971; Nerlove, Munroe and Munroe 1971). Although I did not perform the cognitive tests of Nerlove and her colleagues, my observations of the children's everyday behaviours in Sudan revealed a high degree of self-managed sequencing, as well as a great deal of spatial autonomy. The work of Nerlove and her colleagues (1974) suggests that these in themselves are 'natural indicators' of cognitive ability. In Howa the work and play activities of children lent themselves to and called for the development of a range of cognitive abilities. I noticed few sex- or status-related differences in these opportunities. Children in Howa were essentially equally free to develop these abilities in the course of their everyday activities, however these might differ, at least through middle childhood when, as I have indicated, the options for girls began to narrow and when those (boys) who continued their formal education had to leave the village altogether.

Children in urban areas of many industrialized Western countries have far fewer opportunities for such 'grounded' learning and free exploration. Their access to the outdoor environment is limited largely by parental concerns for their safety – both physical and psychosocial. These concerns have arisen in large measure due to the deterioration of the urban environment brought about by capital disinvestment at the same time as fewer women or other family members are available to care for children when they are not in school or other education programmes. There have been dramatic shifts in household composition over the last two decades in the United States, for example. The number of single-parent families has grown steadily and women increasingly have left home-based care-giving roles for the paid labour force out of economic necessity and/or in response to changing socio-economic relations of production and reproduction. Under these circumstances, safe access to the outdoor environment in the United States is becoming a class privilege, available to children whose parents can afford childcare, safer living environments and/or special programmes. Even then, these options rarely allow the kind of unencumbered and autonomous access that children in Sudan had to the outdoor environment. The erosion of children's autonomy and outdoor experience in industrialized Western cities marks not only a deterioration in the quality of children's lives, but an arena of deskilling with long-term implications both for the children and for the society as a whole (cf. Hart 1979, 1986; Carbonara-Moscati 1985; Torell and Biel 1985; Katz 1991b).

While it is easy for Western eyes to see the limits to mobility imposed by purdah, or the foreclosing of some of life's choices without a formal education, closer scrutiny of the lives of children in our in-dustrialized cities reveals deep impoverishment of experience and a staggering disregard for its consequences, except insofar as they impinge on middle-class concerns. The comparison of girls' and boys' lives in these diverse settings is intended to point to the need for change in both.

CHILDHOOD CITY

Children are curious about the life of their communities. They are intrigued by the activities of their elders and throughout childhood experiment with fitting themselves into the social fabric around them. In places like Howa children have easy access to much of this life. Their lives and those of their elders take place in sociophysical settings that are largely overlapping. Children not only witness but participate in much of the work undertaken by members of their households. The sociophysical domains of children and adults are more discrete in industrialized nations, particularly in urban areas where children's opportunities to observe the range of work and other activities of their elders are severely restricted. In these areas children learn a great deal about the comings and goings of their community by playing in the streets and other public spaces that give them visual access to its everyday life, even as these sites place their activities under the gaze of others.

As urban anonymity, traffic, crime and disinvestment increase, these spaces become more hazardous for children. Parents are loath to allow their children to play outdoors without adult supervision. At the same time, parents are increasingly unavailable to offer their children direct supervision, because of the rise in single-parent and two-earner households. 'Latchkey children' are often isolated indoors for most of their after-school hours. Unable to play with friends or to explore on their own, these children lose important opportunities for social, physical and cognitive development (cf. Erikson 1963; Bruner and Connolly 1974; Bronfenbrenner 1979). These problems are not evenly distributed. Studies have revealed a geographic variation in the experiences of latchkey children, indicating that those in cities are at higher risk of psychological harm. Urban latchkey children have been found to have more problems at school, poorer self-concepts and social relationships, and greater fearfulness when left alone than children in self-care in suburban and rural areas (Robinson, Coleman and Rowland 1986: 4).

Girls are particularly restricted in this regard. All studies of children's outdoor play in industrialized settings have found significant differences between girls' and boys' autonomy, the extent of their free range area, and the ways in which they use space (e.g. Tindal 1971; Anderson and Tindal 1972; Saegert and Hart 1978; Hart 1979; Bjorklid 1985; Blakeley 1987; Matthews 1987). Hart (1979), for example, found not only that girls' home range was smaller than that of boys among all socio-economic groups in a small town in Vermont, but that girls were not encouraged to explore and manipulate their environment to the same extent as the boys in their community. In Stockholm, Bjorklid (1985) found that boys were given more territorial freedom at an earlier age than girls. Noting that most of the outdoor spaces provided in urban areas, particularly those abutting high-rise housing, are flat open spaces preferred and dominated by boys, she found that girls' range was even more constrained in areas of high-rise buildings.[7] Drawing on her study of two housing estates in Stockholm, Bjorklid suggested that girls in the city lost out in three ways; there were few outdoor places for the quieter play activities that they preferred, such as dramatic play, sitting games and socializing; these activities were difficult to undertake outdoors in the cold winter months; and the skills developed in the course of boys' team sports – which predominated in the open spaces of the neighbourhood – were more highly valued by the wider adult society than those developed in the course of the girls' outdoor activities.

In a pioneering study of spatial range among Black children in urban and suburban Baltimore, Maryland, Margaret Tindal (1971) found similar disparities in boys' and girls' experiences. Working with children in the second and fourth grades, Tindal found not only that boys had a greater home range than girls in both age groups, but that the range of the second grade boys exceeded that of the fourth grade girls. Among other things, she found that the boys were more mobile than girls their own age and older, due largely to cycling, and that they had more 'activity nodes' than the girls, who were more restricted in the places that they were allowed to go. Girls were explicitly forbidden to go to alleys, vacant houses and empty lots, for example, and fears instilled at least partially by their elders held them back even further. Tindal's work suggested that fear circumscribed children's worlds, especially at younger ages; and that this tendency was exacerbated in urban areas. While boys have rules that restrict them from going too far or to certain kinds of places, there appears to be greater laxity in the enforcement of these rules for boys than for girls. As Hart (1979) discovered, boys are *expected* to break rules, and the fact that they are not punished when they do, unless 'something happens', must only encourage them in this behaviour. Boys will be boys, after all, and it appears that this is precisely the point – through their socialization which encourages environmental exploration, manipulation and risk-taking, boys develop skills and attributes valued in male-dominated capitalist industrial societies.

Fear, danger and children's safety rattle around at the root of children's geographies in urban settings. In a study of parents' conceptions of children's safety in neighbourhood play settings in New York City, Kim Blakely (1987) found that some parents made a distinction between environmental phenomena that were 'unsafe' and those that were 'dangerous'. The former applied to physical hazards, while the latter referred to social threats to their children's well-being. This distinction points to the barbarous heart of the matter – in the United States a child is injured or killed by guns every 36 minutes; 135,000 children bring guns to school every day, and every 47 seconds a child is neglected or abandoned (Edelman 1991: 6). Boys and girls are sexually molested and physically abused not just by strangers in the anonymous spaces of the city, but at home, in daycare centres, in youth groups. In UNICEF terminology, these are children in 'especially difficult circumstances', suffering in ways as horrific as their peers in the Third World who are refugees, street children, living under war, malnourished, ill. Of course in industrialized countries these threats are not equally distributed, and a disproportionate share of children who are murdered, injured through violence or sexually molested by unrelated adults in the US, are poor, Black or Latin American. Likewise, these children are the disproportionate victims of less dramatic but more widespread and pernicious assaults to their futures, such as the lack of decent housing, education and healthcare. According to Marian Wright Edelman, founder and president of the Children's Defense Fund, 100,000 children in the United States are homeless each night and every 53 minutes a child dies from poverty (Edelman 1991). Through these mind-, body- and spirit-eroding means, children are deprived of their birthright – the means to partake of the social, political, cultural and economic inheritance of their culture – and so are disadvantaged and deskilled.

LOST SPACES OF KNOWING

The distinction between 'dangerous' and 'unsafe' environments points to common bonds in the experiences of girls and boys in Sudan and the urban United States, where many of the restrictions on children's movements were concerned primarily with access to girls' bodies. These patterns reflect the codes of control that define female experience in most of the world. While purdah is recognized as an explicit means of effecting this control, equally powerful codes – implicit and explicit – limit female access to and control over space in the heart of the advanced capitalist industrialized world as well.

As noted above, all studies of children's outdoor experience in Western industrialized settings – rural, suburban and urban – reveal that boys are allowed greater spatial freedom and range from an earlier age than girls. Yet boys are injured with greater frequency than girls from all manner of 'unsafe' elements in the environment – traffic, falls, poor play equipment and the like, as well as from accidents incurred in the 'normal' course of play. However concerned parents are about physical injury, it does not appear to motivate significant restrictions on their children's movements.

Fears concerning the social dangers to which children are exposed, however, seem to impel parents to limit children's, and particularly girls', autonomy outdoors. The data suggest that, at least in the US, boys are the more common victims of social dangers as well. Even as boys suffer more frequent sexual and physical abuse than girls, and comprise the majority of adolescent prostitutes, parents continue to act as if their sons were immune to these social dangers (Smith 1991). Without minimizing the real and present dangers to which children outdoors alone in the US are exposed, it appears, at least in part, that parents are responding more to the horrifying and headline-grabbing nature of social crimes against children than their actual incidence. These perceptions lead parents to exercise greater precaution – especially concerning their daughters' movements – than these events might warrant. But that is the nature of risk perception (e.g. Kates 1977; Fischhoff *et al.* 1981; Douglas and Wildavsky 1982; Kirby 1990), and parents make choices and rules for their children within a risk calculus that is as much a function of fear as of fact, of statistical reality (whatever that is) as of social custom, of emotion as of economics. The result is that girls in urban industrialized areas are restricted from exploring and manipulating the outdoor environment perhaps even more so than girls in ostensibly more restrictive societies such as Islamic Sudan. I am, of course, mindful of the counter-argument that it is because parents keep their children indoors or under adult supervision that social crimes against them are not more prevalent. To a certain extent this is surely true. My argument here is that: 1. parents do not restrict their children's access to the outdoors to protect them from 'unsafe' environments as much as from 'dangerous' ones; 2. childhood injuries from 'unsafe' environments exceed those from 'dangerous' ones in number and in frequency; and 3. girls are perceived to be at disproportionate risk from the latter, while boys are apparently at disproportionate risk from both. The net result is that boys have greater access to and control over the outdoor environment than girls, despite the risks it poses to them; while perceptions of the less common but more violating threats to girls limit their control over space even in their own communities.

The loss of environmental experience has serious ramifications for all children. The life course implications of girls' restricted worlds are especially severe. I referred above to the demonstrated links between autonomous and self-directed sequences of behaviour and the development of analytic ability, and suggested that autonomous exploration and manipulation of the outdoor environment was an important arena wherein children 'self-direct' the sequencing of their behaviour (cf. Nerlove *et al.* 1971, 1974; Hart 1979). Developmental, environmental and cognitive psychologists, as well as anthropologists and geographers, have discovered that boys tend to exceed girls in spatial and mathematical ability, that their cognitive maps and mapping skills are more elaborated and that they are more skilled at related analytic operations (e.g. Nerlove, Munroe and Munroe 1971; Saegert and Hart 1978;

Harris 1981; Liben 1981). While there is some suggestion that these differences can be explained partially by neurological differences that are historically and contextually constructed between males and females (Harris 1981),[8] there is compelling evidence of the dramatic role played by socialization and experience, and it is clear that physiological explanations are partial at best. (See Self *et al.* 1991 for an excellent review of the literature.) My concern is not only that the different spatial experiences of male and female children favour greater development of particular skills and abilities among boys than girls, but that, because of unequal social and economic relations, certain girls may be deprived of the chance to develop these skills at all. The spatial patterns of many urban girls contribute to their deskilling (cf. Valentine 1989 and Pain 1991 who address the ways fear and crime circumscribe women's movements and reinforce patriarchal relations).

Imbalances in the development of spatial abilities may limit affected girls' interest and chances in a range of fields, among them geography, architecture, engineering, mathematics, chemistry and physics (cf. Harris 1981). In the Sudan case I have suggested that restrictions on girls' mobility and socialization may result in their inadequate preparation to be agriculturalists at a time when that role is increasingly thrust upon them due to male outmigration (cf. Katz 1991a). Under both sets of conditions – whether in the United States or Sudan – the limited development of particular skills among females may result not only in diminished life chances but in productivity declines that affect the society as a whole.

CONCLUSION

This chapter has attempted to demonstrate the nature of restrictions on female mobility in two distinct settings, exploring some of the life course implications of these spatial patterns. In examining these two cases against each other, what was striking for me was that despite obvious (potentially blinding) surface differences, there was so much in common in female experience and its meaning for the control over and production of space. Notable among the commonalities in these two diverse settings – rural Sudan and urban United States – was 1. how the particular configuration of the household affects children's spatial experience and how shifts in this configuration can have dramatic impacts upon children's access to the environment; 2. how restrictions upon girls' mobility was rooted in social codes which control access to the female body; and 3. how girls' relative lack of experience with the environment compared with boys' was a form of deskilling with enduring implications for them and the society as a whole.

The differences were equally striking, and depressing in their implications for contemporary urban children. Part of my intent in exploring these two cases was to 'de-essentialize' purdah as a social practice, by showing not only that it was deployed unevenly in Islamic areas, but that in the West female mobility was also restricted, perhaps more perniciously because it is masked by an ideology of equality. In doing this, I surprised myself. My work suggests that children in rural Sudan had a more extensive immediate world than their counterparts in many urban areas of the United States, and that, unlike the US, access to and control over space was not sharply divided by class or gender. Without romanticizing the experiences of children in rural Sudan, it appears that the variation and range in what they were able to encounter in their everyday lives enabled them not only to acquire an extensive knowledge of the surrounding environment, but their freedom to explore and manipulate this environment endowed them with learning skills.

While in Sudan the sociospatial patterns of most females' adulthoods did not allow the development of much of this knowledge, it is possible that their autonomous learning skills will empower them in the face of changed circumstances and novel possibilities. In the case of the United States it is hard to be optimistic. My work and that of others suggests that childhood experience is being eroded in a way that actively deskills children, and particularly girls, reducing their life chances in adulthood. This is aggravated by the structural shrinkage of life chances due to the crises of capitalism that affect the work landscapes of so many children coming of age at the end of the twentieth century. In the midst of my field work in Sudan, one of the most precocious children asked, 'Is America bigger than Howa?' I'm no longer sure of the answer.

5

'HE WON'T LET SHE STRETCH SHE FOOT'

Gender relations in traditional West Indian houseyards

Lydia M. Pulsipher

INTRODUCTION

Women in the former British colonies of the West Indies often live out their lives in quasi-communal domestic units known as 'yards' or 'houseyards',[1] where they interact with males on a variety of levels as they mature from infants to old women. In this culture, derived from a mingling of British and African customs modified by the circumstances of the New World plantation system and more recent modernization, the range of possible male/female relationships is more numerous than and different from those in most other traditional as well as modernized societies. Outside observers have frequently misunderstood West Indian gender relationships, not only overlooking non-mating male/female interaction, but also often judging even mating relationships harshly, according to alien ethical standards.

In this chapter, I interpret the experiences of a number of women now in their sixties, seventies and eighties, showing the role that gender has played and continues to play in their lives; and I argue that throughout the life course West Indian women, in their capacities as aunts, cousins, grandmothers, sisters, mothers, friends and lovers, not only have a rich repertoire of possible relationships with males of all ages, but also that in their typical interactions with males they enjoy considerable power and autonomy.

THE HISTORICAL AND SOCIAL CONTEXT

Although Columbus passed by and named many of the islands of the Eastern Caribbean archipelago in 1493, the Spanish were not interested in these tiny islands. For more than a hundred years the Native Americans, who had settled the islands as long as 2,300 years ago, continued to occupy them, disturbed only occasionally by the Old World invasion. Then, after 1620, British, French and Dutch colonists quickly imposed a new European order on the tropical island landscapes. The land was cleared, divided into owned parcels and planted with cash crops – first cotton, tobacco and indigo and after 1650, increasingly sugar cane. Hundreds and then thousands of labourers were imported to work on the plantations. On the British-held islands Irish and Scots indentured servants dominated this labour force in the 1600s; but by 1730, on nearly all Eastern Caribbean islands, the overwhelming majority of labour was of African origin (Bridenbaugh and Bridenbaugh 1972; Dunn 1972).

Today the culture of the Eastern Caribbean is a blend of primarily African and European traditions. The Eurocentric plantation slavery system imposed an authoritarian, racially-indexed social order which sought to keep the majority, those with any African genetic heritage, in a subservient position. While this system long deprived the African Caribbean people of political power and imposed severe restrictions on access to resources, freedom of movement and employment (even as recently as the 1950s), because of their numbers, and their creativity in designing strategies for subverting the repression of plantation and post-plantation society, the African slaves and their descendants asserted a strong influence on the culture of the region, the extent of which is only now being fully recognized.

Chief among their contributions is the organization of domestic life: the spatial organization, the material culture, family organization, child-rearing customs, value systems, gender relationships and food production and distribution. It should be noted that this African Caribbean complex of features, which came to be the traditional culture of the region, is now itself undergoing rapid change due to the influences of modernization.

THE CARIBBEAN HOUSEYARD

The essential unit of Caribbean traditional domestic culture is a residential arrangement often called a yard or houseyard. Found in both rural (Pulsipher 1992) and urban (Brodber 1975; Austin 1974) settings, these domestic spaces embody a distinctive communal way of life that was once widespread throughout African America from Brazil to South Carolina and beyond.[2]

The houseyard as a social institution is grounded in the slavery era. The historical literature contains a number of references to the domestic spaces of slaves and freedmen, describing them as clusters of houses and outbuildings around central activity areas, interspersed with economic plants and animals and inhabited by people linked through kinship and

Plate 5.1 Typical houseyard in Montserrat

friendship (Wentworth 1834; see also Olwig 1985; Craton 1978; Handler
and Lange 1978). Writing in Jamaica in 1818 one author described a
slave habitation:

> In some instances whole families reside within one enclosure: They
> have separate houses, but only one gate. In the centre of this family
> village the house of the principal among them is generally placed,
> and is in general very superior to the others.
>
> (Anon. cited in Higman 1974)

Another writer in Jamaica in 1825 described a five-dwelling houseyard
that contained the eighteen-member extended family of an African-
born matriarch, Bessie Gardner (Higman 1974). Furthermore, archaeology in slave villages in Montserrat and Jamaica has confirmed the
general pattern of a cluster of dwellings interspersed with plants and
various activity areas (Howson 1987; Armstrong 1990).

In the modern Caribbean, although the size and morphology of
houseyards can vary greatly, they usually range in size between a
sixteenth and half an acre, and are commonly occupied by two or more
small wooden chattel houses,[3] or their modern concrete replacements.
The houses in yards vary in size, but a typical house measures 6 feet by
12 or 10 feet by 20 and usually contains two rooms. Often the original

109

house will have one or more additions, or two houses will be joined to create more interior space. In addition to houses the yard will contain other structures such as detached kitchens, ovens, tool sheds, animal pens, work benches, showers, privies and laundry facilities; and the entire complex will be arrayed with a variety of useful plants: coconut and fruit trees, vegetables, ornamentals, medicinals and other plants that have a practical use. A houseyard is usually distinguished from adjoining property by some visual boundary: a fence, a row of croton bushes, or bamboo or agave, or a soil bank (Pulsipher 1986).

In most yards, one of the residents, often the oldest woman, owns the land or has informal rights to it due to long-standing family or employment connections. Residence in a yard may be attained through a variety of circumstances ranging from close kinship with the head of the yard to a purely business transaction wherein a yard resident pays rent and is only peripherally involved in yard activities. A resident in a yard may own her/his actual house but rent the land on which it stands. This situation was more common in the days when all houses were of wooden construction and could be easily moved. Then the array of houses in a yard could be added to or rearranged at will.

The relationships of those who occupy a particular house in a yard are varied, but rarely are they nuclear families. Some possibilities include: a mother and her young children who are occasionally visited by one of their fathers (usually that of the youngest); an older woman and several of her adult children (sometimes male, but more often female) and their children; adult (uterine) siblings,[4] or cousins, and their children; a single adult with 'borrowed' children;[5] a lone single adult; a sexually cohabiting couple with only their common children, or with their common as well as 'outside' children;[6] or an elderly couple living alone or with several grandchildren belonging to both, or one of them. It should be noted that the simple nuclear family (married parents with children all born in wedlock), so often identified as a Caribbean norm or ideal (Smith 1988: 180), is most unusual and never emerged in my sample.[7] Although yards typically are made up of houses, outbuildings, domestic plants and animals and multiple residents, what really distinguishes these domestic units is the spatial patterning of life. The houses themselves are used only for sleeping, for keeping clothes and valued possessions and for sheltering from intermittent rain. The yard is where most of life is lived. Here is where people prepare food, eat, tend animals, care for and process plants, bathe, wash clothes and conduct most social transactions (Pulsipher 1992).

Although most houseyards have several residents – some as many as twenty or thirty – it is possible for a houseyard to have only one occupant. This situation is recognized as abnormal, and untenable, because the essence of yard life is the community; but with modernization,

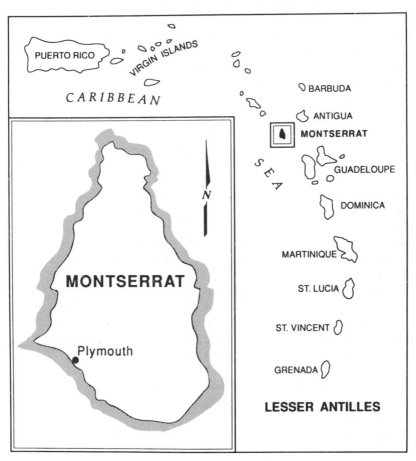

Figure 5.1 Location of Montserrat

and especially massive migration, often one elderly person is left alone in a yard that once rollicked with life.

In this chapter I concentrate on the experiences of Eastern Caribbean women recorded over the period from 1973 to 1991; but the most concentrated gathering of information took place in eleven houseyards on the island of Montserrat between 1985 and 1987 when I was conducting an ethno-archaeological study of domestic life in conjunction with the long-term interdisciplinary study of Galways Plantation on that island.[8] This study required detailed observations of the material culture of houseyards and of the human interactions that took place within these settings.

The research design called for noting gender relationships, but my North American cultural bias led me at first to expect that the significant

111

gender relationships would be those between mates and between parents and children. It soon became clear that these were only two of many cross-gender interactions that took place on a daily basis; and that in terms of time spent and psychic energy invested, these other gender relationships were at least as important as mating and parenting experiences.

THE TYPICAL LIFE COURSE OF FEMALES IN HOUSEYARDS

Females, more often than males, can spend most of their entire lives in one yard but their roles in that space change radically over the life course due to such factors as maturation, mating, motherhood, the birth and death of kin, jobs outside the yard, migration, possible marriage (often in middle age) and finally ageing.

Toddlers of both genders learn early to make contributions to yard life, first as entertaining companions to all, then as minor assistants, perhaps fetching objects for the elderly, or feeding scraps to the chickens. From the age of three both boys and girls alike help in caring for younger children; and this childcare role continues for both genders into adulthood. Nonetheless, as they mature, girls are steered more and more into particular domestic tasks: laundry, house and yard-sweeping, and food preparation – activities that are likely to keep them in the yard. Boys, while they may help with (and be quite skilled at) such tasks, are more likely to be found outside the yard, tending animals, making trips to the store, going into the mountains to tend provision gardens (Pulsipher 1992).

Although this chapter concentrates on traditional Caribbean gender relationships other than mating, it is important to understand first how mating customs and the resultant family organization fit into the life course of Caribbean women. To some extent, it is the mating system and the female autonomy it facilitates that sets the stage for the variety of gender relationships that are possible in the Caribbean.

As girls reach their early teens, special care is taken to guard them against unsupervised contact with non-family males of any age over puberty, including even elderly men. The goal is not entirely to prevent sexual activity, but rather to make sure that the potential mate is suitable. Suitability is determined by the girl's preference and by evidence that the male has economic prospects, that he respects the girl and will continue to consider her welfare and that of any children that result (Clarke 1966: 105; Jagdeo 1984: 60–4). There is little expectation, on the part of the girl, her family or the mate, that the relationship will be lifelong. The girl will, in most cases, continue to live in her customary yard, with the male only visiting several times a week and taking some of his meals there.

112

Girls may have their first babies as early as age fourteen or fifteen; but often their mothering role is rather limited after the first few weeks because childcare is willingly taken over by older kin or unrelated yard residents, especially females, but also males. Whereas in the past young mothers might have worked as agricultural wage labourers, now they tend to return to school or jobs in the service or industrial sectors of the economy. In fact, the houseyard system, which provides shelter and opportunities to cooperate in childcare, gardening, animal tending, laundry and food preparation, allows young women to be astonishingly independent, despite the early onset of motherhood. This is especially true today for women who can find white-collar jobs. Their wages give them considerable autonomy because, while they contribute to the economic well-being of the yard, perhaps financing new concrete houses or the addition of modern plumbing, they are freed of many domestic chores; meanwhile, their more traditional female relatives are still willing to stay at home and perform labour-intensive domestic duties. As modernization proceeds, extending opportunities for employment outside the yard to more women, this advantage of yard life probably will disappear.

The father in a mating liaison visits the mother and child, providing in-kind and monetary support; the parents may even live together for a time in her yard or his. A second child may result, but just as likely the sexual relationship will dissolve and both will establish new mating alliances. A woman may have one or two, or as many as six or seven mates over the course of her reproductive life, sometimes marrying the last if he happens to be suitable as a life-long companion, or has attractive economic assets. A liaison may be established on the basis of love and attraction, or the explicit and sole intent may be to have a child.

While the tendency of women to have children by several mates is changing as education and employment options for women increase, some educated women with well-established careers may choose to have a child with an acquaintance, and marriage or even long-term involvement may never be a consideration. For a person to go through life without the experience of child-raising, as well as the companionship and support of children as they mature, has been considered a major deprivation since slavery days. It is still expected that a childless woman, or a lonely person whose family has grown, will 'borrow' a child from a sister or friend to raise as her own. Likewise, a man, childless or not, will take a special interest in, and sometimes will take over the actual raising of, one or more of his sister's or a female cousin's children.[9]

As a woman matures, she may migrate to work abroad for several years, sending home remittances to support her children and/or her elderly relations who have been left in the care of family and friends in the houseyard. Or she may stay home and begin to assume more autonomy and responsibility within the yard as an organizer and supervisor of the many domestic activities and as a financial contributor through her wages. Often the middle-aged woman, who by this time may have decided to marry, will move into the house of the elderly yard head, physically taking over the central space but continuing to defer to the authority of the more senior woman in the yard – an aunt, mother or grandmother. Meanwhile the oldest woman will continue to prize whatever autonomy her physical condition will allow, cooking her own meals and caring for herself, earning money through the sale of her goats, chickens, garden produce or handicrafts, taking in laundry or selling processed food like cassava bread or guava jelly. Elderly women take pride in their independence as long as possible, citing recent instances when they contributed food or cash to others in the yard; but when such self-reliance is no longer possible, they will take equal pride in being able to say that their family members are supporting them and that they no longer need to work.

GENDER RELATIONSHIPS IN THE LIVES OF SPECIFIC WOMEN

Whereas it is possible to corroborate statistically some of the observations I have made regarding gender relationships and family organization (for example more than 70 per cent of the children born in the Eastern Caribbean are born to non-married parents, Jagdeo 1984: 5; Smith 1988: 22), most of the nuances of gender relationships can be garnered only from lengthy observation of houseyard life and from the personal accounts of individuals. The following personal histories are three of many that have revealed to me the actual nature and complexity of cross-gender interactions in the Eastern Caribbean.

Nancy Dyett

Nancy Dyett was born in 1906 to Martha Dyett, a young Montserratian woman and Marshall James, an Antiguan. They met in Guadeloupe where both were working temporarily in the cane fields. When the child was three, Martha left Nancy in the care of her grandparents, James and Judy Dyett of St Patricks village, Montserrat, and migrated permanently to Colón, Panama. There Martha worked, like thousands of other West Indians, in jobs related to the development and operation of the Panama Canal. She married and had more children, but she never returned to see her first child; and the father Marshall James never contacted the child again, either. Nancy spent much of her childhood in the company of her maternal grandfather, James, learning from him lore about Galways Mountain above the village and about their family.

114

Plate 5.2 Elderly woman resident of houseyard in
Montserrat

He taught her the skills of tropical gardening, animal tending and
bread baking that she has used to support herself and her family ever
since.

Nancy grew up in a large multi-dwelling houseyard with several
generations of relatives, including her grandparents, eight aunts and
uncles (the siblings of her mother) as well as several great aunts and
uncles and cousins. As she matured, Nancy took over the care of her
grandparents and a series of elderly relatives until their deaths. She
learned to sew and she gardened, raised animals, baked bread and

started two small shops in the yard. From one she sold tinned goods, matches and other small necessities; the other was a rum shop. When one maternal uncle died, leaving her a small inheritance, she began a period of inter-island 'higglering' (bartering or trading), selling vegetables, chickens and other items in Antigua, St Kitts, Guadeloupe and Montserrat. From these activities she eventually saved enough to finance an eight-month tour of the US, Canada and England, visiting friends and relatives.

Throughout her life Nancy had a series of mates, but these are not the relationships with males that she emphasizes in her personal history. Her grandfather, certain friends past and present, her uncles, certain cousins and her 'borrowed' son loom much larger. She says that she could have married many times but really never considered the possibility because it made more sense to rely on herself. That way she could not get saddled with an unfaithful or lazy husband, and could maintain her ability to pick and choose lovers. At eighty-five, Nancy remains an attractive and sensuous woman who enjoys raucous joking relationships with men of all ages as she goes about her animal-tending in the mountains, undertaking occasional wage labour clearing land with her cutlass, travelling to market in Plymouth, walking through the village or sitting outside her door in the large yard that she now occupies quite alone and finds difficult to maintain. Her conversations with male friends reveal that she has long enjoyed their camaraderie. Their jokes span decades of familiarity, referring to both tragic and hilarious events occurring only yesterday or deep in the past. Until his death several years ago, one of her closest companions was a man several decades her junior with whom she occasionally worked on the Galways archaeological site. They looked after each other in little ways; and she reports having a vivid dream about him the night he died.

One of the ways Nancy has supported herself in her old age is by renting out (until it was destroyed by Hurricane Hugo in 1989) a small house in her yard to a middle-aged gentleman who ran a sedate rum shop in it. Customers would leave complimentary drinks for Nancy with the bartender, which she retrieved at her leisure when she was in need of a little company. As she downed her complimentary drink, she had the opportunity to enjoy a few moments of spirited verbal interaction with the mostly younger men who came to the bar.

According to an Obeah woman in St Kitts, although Nancy had several babies inside her, the 'seed' of her various lovers had just never matched up with Nancy's seed so she had never borne any children. Instead she raised a borrowed child whose mother had died leaving her in the care of a godfather, a friend of Nancy's, who could not keep the little girl because his present lover objected. This child then had a baby at the age of 14, so Nancy raised him as well; and it is this 'son' who continues to look after Nancy financially, by paying for her telephone and sending her cash periodically, though he has not returned from England to see her in several decades.

Alice Roach

It is the summer of 1989, the sun setting to the west sends streams of golden light into the tiny lean-to kitchen where ninety-year-old Alice Roach stands over a skillet balanced on a coal pot, which in turn sits on an old weathered plank set across an empty oil-drum. She is frying fish for her fifty-seven-year-old friend, Ned. Ned, who is the sexton for the nearby Anglican church, comes by every evening on the way to ring the church bell. His mother died recently and it was then that he began making the daily visits to Mrs Roach, as he respectfully calls her. She needed his help and he was lonely. He provides her with vegetables from his garden and occasional purchased items like dry milk and chicken from town. He makes charcoal for her to use in cooking, and tends her yard, pruning plants, cutting down weeds. In return, Alice is responsible for cooking for Ned and for repairing his clothes. When she fell out of bed recently, he took her to the doctor, applied the medication prescribed and massaged her bruised body. Alice, a sardonic, irreverent, twinkle-eyed woman, who maintains a lively interest in sex, repeatedly explained that their relationship is entirely platonic. Yet, she jokes about him not being able to control himself much longer. Marriage may become necessary, she says, to quiet the gossip of villagers who think they are too intimate. Ned, a serious fellow, is embarrassed by her joking. He protests that he deeply respects Mrs Roach, and merely enjoys her company and wants to help her to honour the memory of his mother who was her friend.

Alice began life as a servant girl when her mother, a labourer on a remote plantation, sent her at the age of eight to work in town in the houses of merchants and planters. The idea was that she would learn to cook and sew and to read and write, so that she could avoid low-status field labour. At the age of thirty-three, when she had already saved enough to build a small wooden chattel house and was working as the housekeeper of a bachelor plantation manager, Alice decided to get married. Interestingly, the man in question, younger by seven years, was unaware of the plan until he was told by friends one Sunday afternoon that the banns had been read in church that morning. The marriage lasted one year, when it was discovered that he was interested in other women. After this episode, Alice, disheartened by the experience (divorce being quite a disgrace), moved her chattel house to a relative's yard in another village. There, she acquired a young boy to raise as her son. The cost of a child meant that she needed more food and cash.

This need, and the fact that she had a dispute with another resident in the yard, caused her eventually to move her house to rented land near the cotton fields where she share-cropped cotton and raised a subsistence garden and goats. Once the child was grown, in 1960, Alice again decided to marry. This time she chose a new widower, who owned 'a cow and a calf, a donkey and a cutlass', as Alice tells it. Since she owned a house, he came to live with her.

'. . . and I was glad for the company. For twenty years we lived on that cow. Every year it would have a calf, we would milk the mother, sell the milk, raise the calf and then sell it. I hardly ever had to raise cotton, for cash, just a provision garden. But, I did raise cotton anyway, because I wanted to buy some land. But when Mr Roach [she reports that she always called him 'Mr Roach'] got old and sick, we had to sell the cow, because I couldn't control it. I was scared of it. Then I had to raise hot peppers and make charcoal for cash. And I never got the land either. The cotton land manager took my cotton every year and marked it down in a book as payment toward the land. Then he died while on a trip to England; and no one would honour the "tally".'

Mr Roach died in the early 1980s; after several years of looking after herself in conditions of considerable hardship, eventually Ned came into her life. Then, in the autumn of 1989, Hurricane Hugo completely destroyed Mrs Roach's little wooden house. She now lives near the capital, Plymouth, in an apartment in the large modern home of the man she raised; but he has migrated to the United States, leaving her in the care of a diffident young woman, his wife's niece. Ned left Montserrat shortly after Hugo, to stay with family abroad. Only her advanced age keeps Mrs Roach from establishing a new liaison with a male; she constantly speculates on the possibilities.

Margery Tuitt Watkins

It is noon on a summer Sunday in Cork Hill village, sixty-year-old Margery Watkins has just finished helping her mother bake some twenty loaves of bread in the large stone oven that is the centre-piece of their half-acre yard. Sixteen of their relatives, the children and grandchildren of Margery and her sister, Alberta, are busy in and around the five houses, numerous sheds, animal pens and garden plots that make up the yard. Sitting in a clearing under a breadfruit tree, eighteen-year-old Basil has just finished feeding the baby daughter of his cousin, Leonora, and now he is struggling to dress her in a frilly pink dress. Four-year-old Suzanne, the child of his sister, is waiting in line for him to comb her hair. He lives with his cousin and her children in one of his grandmother's houses. They share expenses, she usually cooks for him and he helps her with the children. His brothers and male cousins, several of whom have children of their own living in other yards with their mothers, fulfil similar roles in this, their home yard. Like the young women in the yard, they frequently help with domestic duties and childcare, and some of their wages buy food and necessities for their nieces and nephews and the older women.

Margery has four adult children, the last two of whom were born in the 1960s in England during the ten years she spent working there as a domestic servant. All four have different fathers, with whom they now have little contact. In 1974 Margery married a man thirty years older than herself, with whom she has never lived. At the time they were both working for an elderly, bed-ridden, expatriate resident of the island. They became good friends and began cooperating in several small-scale economic activities, like making and selling ice (from the refrigerator of their employer) and rearing, butchering and selling goats and sheep; but there is some evidence that they also shared connections through the folk belief system, known as Obeah. Both are evasive about why they married and then kept the transaction secret, never living together. He reports that he hoped Margery would care for him as he aged, which has not happened; and he also indicates that they married in part because the idea was ridiculed by their employer, who was never formally told of the marriage and only learned of it through hearsay, years later.

Today, there is little or no contact between the couple. He lives alone in his home village several miles away. Margery, on the other hand, although she refuses to talk about her husband, and gained little financially from the marriage, seems to enjoy the social status that goes with being a woman who married in maturity.

Margery's pride and joy is her son, James, who is now a police officer in Antigua. He sends her money and parcels occasionally and every so often invites her for an extended visit. But, most of the time her male companion is her grandson, Craig, whom she 'borrowed' from her daughter, Leonora, shortly after his birth in the late 1970s. Until Hurricane Hugo his mother lived in the same yard, sharing the house with Basil, her cousin, and her other children; she now lives in a distant village and only occasionally sees Craig, who is in his mid-teens. Margery has always provided for her grandson in every way, seeing that he is well clothed and equipped for school and gets formal religious training at the Anglican church. As he matures, Craig is more and more useful to his grandmother, helping her with her goats, her large provision garden, and making and selling charcoal. Such a cross-generation, cross-gender relationship is not at all unusual, and the closeness of Margery and Craig is commented on with pride by their kin. Mrs Tuitt, the great-grandmother, noted one day that Margery and Craig were so inseparable that he wouldn't even let his granny take a walk without him. As she expressed it in Montserrat dialect, 'He won't let she stretch she foot.'

CONCLUSION

African–Caribbean family and gender relationships have been the subject of speculation by outsiders for more than two hundred years (e.g. Edwards 1793; Higman 1975; Jagdeo 1984; Patterson 1967; Smith 1962; Smith 1963, 1973, 1987, 1988; Stewart 1823; Wentworth 1834; Young 1793). Focusing primarily on mating, observers have been preoccupied with what they usually took to be easy, if not promiscuous, sexual activity; they usually overlooked the fact that African–Caribbean people experience a host of cross-gender relationships over the course of a lifetime, only a few of which are based on sexuality. Being used to familial male dominance in their own cultures, many observers thought they detected residuals of Caribbean male dominance (which some writers, especially missionaries, proposed should be reinforced by religious or civic institutions), or they saw the pathological lack thereof in 'matrilocal' residential patterns and 'matrifocal' domestic relationships. Though few said so explicitly, many left their readers with the impression that the Caribbean family was in a state of disintegration, because women and not men exercised so many choices and were often the heart of family life; or they left the impression that the family was somehow permanently crippled by the plantation-slavery past and that women were the overwhelming victims, since males were not behaving like normative husbands and fathers. In so misperceiving the situation, writers, historical and modern, have slighted both Caribbean males and females. By focusing on absent fathers, they have missed the fact that these same men, whom they define as absent, are devoted uncles, male cousins, brothers and grandfathers who tend the children of their female consanguineous kin when they are young, nurturing and teaching them a host of useful skills. When the children grow up, these men finance trips or business investments for them, and then are themselves cared for by their charges in old age. Males, rather than being dysfunctional in the family, or marginalized, as some have suggested (Greenfield 1966; Patterson 1967; Smith 1962), play out their roles most often in their son/(grand)mother, nephew/aunt, brother/sister, elder sibling, elder cousin and maternal uncle relationships (Pulsipher 1992). Likewise, by focusing on the fact that women have babies outside of wedlock, writers have missed the fact that women often choose autonomy in sexual liaisons and establish instead over their entire life course a series of close, mutually supportive non-sexual relationships with male consanguineous kin and friends of all ages. The significance of a particular male/female relationship may wax and wane, but at any age an individual is likely to have one or more close cross-gender associations. The stories of Nancy Dyett, Alice Roach and Margery Watkins and the males in their lives are not aberrations, they are the norm. Furthermore, a reading of historical documents suggests that this is how gender roles have been played out in the Caribbean for hundreds of years. Yet, it is not certain that this pattern will continue for much longer. Modernization appears to be modifying, if not obliterating the houseyard form of domestic organization. Changes in spatial relationships almost always go hand in hand with changes in human relationships and these three case studies indicate that migration of the young, especially, may mean that elderly women are deprived of all support, notably that of their male consanguineal kin and friends.

6

WOMEN, WORK AND THE LIFE COURSE IN THE RURAL CARIBBEAN

Janet Momsen

The life courses of rural women in the English-speaking Caribbean differ significantly from those experienced in other parts of the world. The contemporary Caribbean pattern has its roots in slavery and the plantation system. During the period of slavery, women were expected to perform much of the heaviest labour in the sugar cane fields, with little respite for pregnancy, childbirth or breast-feeding. The restrictions on marriage enforced by slave owners weakened the conjugal ties, while leaving intact the mother–child bond. After full emancipation of the slaves in 1838, male ex-slaves had the freedom to migrate in search of new economic opportunities but former female slaves found their freedom of movement limited by childcare responsibilities. Sex-specific migration encouraged the further expansion of female-headed house-holds. Today in the Eastern Caribbean, 35 per cent of households have female heads and about one-third of small farms are run by women (Momsen 1988a: 88). Some two-thirds of children are born out of wedlock. Marriage is often entered into only after the reproductive life stage is complete; it is seen as providing respectability for older women, rather than economic support and legal status for children.

Caribbean women have been able to raise children without resident fathers because of a family system of matrilocal households based on house-yards. Often grandmothers will undertake childcare, while their daughters take on paid work. Female economic autonomy has been assisted by relatively non-discriminatory access to land, the central productive resource. In the Commonwealth Caribbean both men and women have inherited land and have had equal access to family land. Although in some areas land is now less important for its agricultural productivity *per se*, it remains an important symbol of status, security and autonomy (Besson 1987). Land reform, which in Latin America has tended to reduce women's access to land, has not had this detrimental effect in the Caribbean (Momsen 1987). However, farms run by women tend to be smaller and less accessible, and to have fewer resources than male-operated farms (Henshall 1981).

In addition to the physical resource of land, women have had access to educational resources for a long period. Women value education and often struggle to maximize it for their children as the best means of ensuring their upward economic mobility. There are clear cohort differences in educational attainment. Older women rarely have more than primary education; younger women, however, are not only better educated than their mothers, but are becoming more qualified than men. For example, in the British dependent territory of Montserrat, a higher proportion of women than men under thirty-four years of age in 1980 had completed secondary education (Caricom n.d.a: 70). The larger and independent nation of Barbados illustrates the sex-specific nature of recent changes in educational attainment: in 1980 14.5 per cent of adult women but only 12.2 per cent of men, as compared to 6.3 per cent of women and 7.0 per cent of men in 1970, had passed some secondary school leaving examinations. It is also noticeable that although in the Barbadian population more adult men than women hold university degrees, for the cohort aged fifteen to twenty-four in 1980, more women than men have a university education (Caricom n.d.b: 30–1). In many cases where men have sought economic opportunity through migration, women have achieved it through education. This is particularly true in those islands, such as Montserrat and Nevis, which have had high levels of male emigration.

In this chapter I focus on the influence of life stage on the multiplicity of Caribbean women's economic roles, with special reference to the poor, semi-proletarianized, small-farming population. I draw primarily on data I have collected during field research in the region, especially in Barbados, Nevis, St Vincent and Montserrat. Changes in women's productive roles will be considered in terms of occupation, income and hours of work. Variation with levels of development will be demonstrated by using comparative data from the lesser and more developed islands of the Eastern Caribbean and by considering time series data for specific islands wherever possible.

WORK AND THE LIFE COURSE

Caribbean women have always engaged in productive and reproductive work. Under slavery, it is almost tautologous to say that economic activity rates for men and women were at the maximum. With emancipation and the introduction of individual choice, many women left the formal workforce to concentrate on home duties. Within a few years, however, male migration and the consequent increase in female-headed households forced many women back into plantation labour in order to feed their children. Between 1891 and 1921 in Barbados, during a period of high male migration to Cuba and Panama, women constituted over 60 per cent of the island's labour force, compared to 45 per cent in 1980. Consideration of the data from Barbados provides evidence for a possible U-shaped curve of female activity, although the upturn of the curve has only occurred since 1970, after an eighty-year decline (Table 6.1). Divergence between male and female rates of economic activity became apparent in 1946 and Lewis (1950: 3) noted similar trends on other Caribbean islands as 'women retired from employment into the home'. The first sign of a reversal of this trend came in the 1980 census figures for the Commonwealth Caribbean (Reddock 1989: 54).

Table 6.1 General worker rates in Barbados, 1891–1980

| | General worker rate | |
Year	Males	Females
1891	79.6	77.2
1911	77.9	76.5
1921	79.6	76.1
1946	72.6	48.8
1960	67.8	36.6
1970	62.5	34.1
1980	64.5	45.7

Source: Adapted from Massiah (1984), p. 23 and Caricom (n.d.a), Tables B1 and C2.

Several factors account for the recent changes. Growth in the proportion of the population which is of working age accounts for an increase in economic activity for both sexes in 1980, and that year recorded the highest ratio of women to men workers since 1921. In part, the increase in economic activity was greater during the 1970s for women than for men because of declining fertility rates and improved female educational levels. Changes in the economy also influenced the new pattern. As Barbados moved from an agricultural to a diversified economy, light industry and tourism expanded the employment opportunities for women (Massiah 1984). There is some evidence of similar trends in other Eastern Caribbean territories.

Women workers throughout the world display distinct regional patterns of age-related economic activity, although these spatial differences are not apparent for men (Figure 6.1). Within the Anglophone Caribbean there are inter-island variations (Massiah 1984): Barbados, Jamaica and Montserrat in 1970 had employment profiles for women in which maximum rates occurred at the age of twenty, while Guyana recorded a maximum rate at age forty-five and St Vincent at age thirty-five. The remaining territories record a bimodal pattern, with a primary peak at age twenty to twenty-five and a secondary peak after thirty-five. The shape of the curve for Barbados has changed quite markedly over time: in 1946 maximum participation occurred at age twenty, followed by a

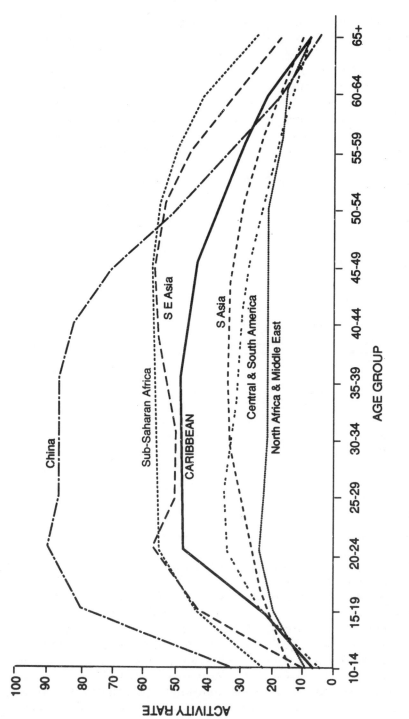

Figure 6.1 World regional economic activity patterns of men and women by age group

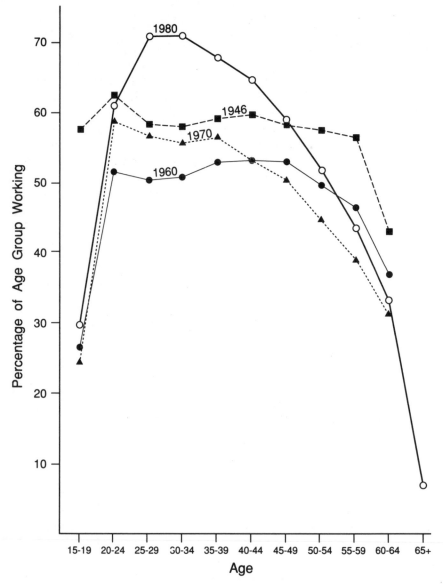

Figure 6.2 Barbados female age-specific worker rates, 1946–80

secondary peak at age forty; by 1960 the maximum rate was not achieved until age forty-five, while by 1970 and 1980 there was a single peak at age twenty, followed by a gradual decline, steepening after age thirty-five in 1970 and age forty-five in 1980 (Figure 6.2). Massiah (1984: 32) suggests that these changes may reflect contrasting patterns of employment opportunities, associated with different stages of economic development.

126

Globally, employment rates for women peak at a lower age than for men, and decline slowly from the age of thirty-five. The decline in female participation rates in the workforce after the age of forty-five, as shown in the 1960 Barbados Population Census, is almost entirely explained by women leaving the workforce to take up home duties. There is anecdotal evidence that some women emigrate at this stage in their life, usually to join adult children working abroad. Massiah's (1984) use of cohort analysis for Barbados shows that economic activity rates have declined most markedly for young women. This is directly attributable to increased secondary school attendance. Her suggestion that the increased rates of participation in the workforce shown by women aged between twenty and thirty-five in 1970 implies imminent changes in the activity of cohorts reaching these critical ages within the next decade is borne out by the results of the 1980 census (Table 6.2) (Massiah 1984: 30).

Table 6.2 Economic activity rates for men and women in 1980

| Age | Barbados | | Age | Montserrat | |
| | Men | Women | | Men | Women |
	(percentage)			(percentage)	
15–19	76.5	62.9	15–19	85	73
20–24	90.0	77.4	20–34	93	75
25–44	95.4	74.9	35–49	93	64
45–64	87.4	48.4	50–64	73	30
65+	23.6	7.1	65+	36	9

Sources: Caricom (n.d.a.), Table C4 and (n.d.b) calculated from Table 2.1.

Spatial as well as temporal variations in the female economic activity rates are also notable within each island. In Barbados in 1960 it was the northern parishes, that is, the most rural, which exhibited the highest female activity levels. By 1970 the highest rates were found in the southern parishes surrounding the capital city and were probably associated with employment in tourism. In 1980 the link with tourism was confirmed, as the highest female activity rates were recorded for the parishes of Saint James and Christchurch, which contained the major concentrations of tourist accommodation. This is quite different from the pattern found in Jamaica, where female employment increases with distance away from metropolitan areas. The smaller size of Barbados and the ease of communication may explain the differences. In the rural parishes of Barbados, participation in the paid labour force is characterized by low levels at younger ages and relatively high levels between ages twenty-five and forty-four, followed by only a gradual decrease with age, closely resembling the patterns found for rural areas in the United States, Canada and England (Momsen 1988b). In Montserrat the

Table 6.3 Occupations of women by age, Montserrat, 1980

	15–19	20–34	Age 35–49 (percentages)	50–64	65+
Professional /Technical	5.9	60.9	20.3	10.2	2.7
Administrative /Managerial	5.7	22.9	48.6	22.9	–
Clerical	18.3	59.8	16.9	3.4	1.7
Sales	22.0	38.2	15.6	15.0	9.3
Services	7.7	45.4	28.4	15.6	2.9
Agriculture	2.9	20.1	20.9	35.3	20.9
Production	18.3	53.6	12.5	7.6	3.6

Source: Caricom (n.d.a).

concentration of older women in agriculture is even more marked (Table 6.3).

SMALL-SCALE FARMERS

In order to understand the processes occurring in these small transitional economies, and the ways in which they affect women's life courses, my focus will now narrow to small-scale farmers in Nevis, Montserrat and Barbados. On these small islands the distinctiveness of and distance between rural and urban areas is much less than in most parts of the world. Farm families are often involved in urban employment, even while they cultivate the land. Women on these farms may have several roles in addition to their domestic duties: they may be the decision-makers on the farm, or they may assist the farm operator by providing labour; they may buy and sell farm produce locally or on a neighbouring island, or they may work in a non-farm job. Individuals may fulfil all these roles during their lifetime, but they are more likely to do so sequentially than simultaneously. Occupational multiplicity is far less common among women farmers than among men. Conversely, women are more likely than men to depend for a large proportion of their income on remittances, usually from family members working abroad (Momsen 1986).

Within these broad patterns life stage is influential. Fewer women than men have off-farm jobs and they are most likely to be involved in this type of work between the ages of thirty-five and forty-nine in Nevis, Montserrat and St Vincent, though the peak age for this type of employment comes earlier in Barbados (Tables 6.4 and 6.5). In St Vincent and Nevis, two of the least developed islands, lower percentages of women than men hold skilled off-farm jobs, and farming remains important for them as a source of income and to provide food for the

Table 6.4 Types of employment of male and female small-scale farmers in Barbados, Nevis and St Vincent

Age	Male			Female		
	Work on own farm only	Off-farm work Skilled	Unskilled (percentage)	Work on own farm only	Off-farm work Skilled	Unskilled
		BARBADOS				
		N=87			N=39	
20–34	9	59	32	14	43	43
35–49	25	56	19	38	50	12
50–64	15	50	35	70	15	15
65+	79	7	14	100	0	0
		NEVIS				
		N=64			N=34	
20–34	0	0	100	100	0	0
35–49	40	40	20	50	0	50
50–64	46	15	39	83	0	17
65+	70	15	15	100	0	0
		ST VINCENT				
		N=33			N=27	
20–34	75	0	25	83	0	17
35–49	43	36	21	60	10	30
50–64	80	20	0	75	0	25
65+	100	0	0	33	0	67

Sources: Barbados – fieldwork 1987; Nevis – fieldwork 1979; St Vincent – WAND, Barbados. Mss. survey, 1981.

Table 6.5 Gender differences in types of employment of small-scale farmers, Montserrat, 1973 and 1983

Age	Male			Female		
	Work on own farm only	Off-farm work Skilled	Unskilled (percentage)	Work on own farm only	Off-farm work Skilled	Unskilled
		MONTSERRAT – 1973				
		N=45			N=20	
15–34	0	100	0	100	0	0
35–49	40	40	20	50	0	50
50–64	46	15	39	83	0	17
65+	70	15	15	100	0	0
		MONTSERRAT – 1983				
		N=79			N=36	
15–34	27	53	20	100	0	0
35–49	0	91	9	50	0	50
50–64	45	14	41	65	6	29
65+	71	10	19	93	0	7

Sources: Montserrat 1973 – fieldwork; Montserrat 1983 Agricultural Census questionnaires.

family. Particularly in St Vincent, the poorest island, older women farmers are very dependent on supplementary income from working as 'higglers' (traders) (Table 6.4). In more developed Barbados there is little gender difference in skill levels of off-farm employment (Table 6.4), and on the farm we see an increase in 'hobby farming' which allows those with full-time off-farm jobs to enjoy their own fresh produce. Montserrat appears to be transitional between these two groups of islands. It experienced considerable development and increased employment opportunities for young women in the light industry and services between 1973 and 1983 but this has had little effect on the older women who remain in the small farming sector (Table 6.5). Clearly women entering these new jobs no longer farm even part-time.

Male farmers in the less developed islands tend to work as self-employed craftsmen such as carpenters, masons or as fishermen; comparable jobs for women are bakers or seamstresses. Most women, however, hold lower-skilled jobs as domestic servants, agricultural labourers or small shopkeepers. In Barbados both men and women farmers have been able to find full-time jobs in the civil service or teaching; where they held less skilled jobs, these were quite likely to be in tourism rather than agriculture.

The role of agriculture at different ages is seen even more clearly if we consider its contribution to total cash income on these small farms (Table 6.6). Income from sale of farm produce was most important for male-headed farm households in Barbados and Nevis when the farmer was between the ages of thirty-five and forty-nine. This was the life stage at which the households surveyed contained the maximum number of children, an average of 1.4 in Barbados and 4.6 in Nevis. For Barbadian female-headed households, greatest dependence on the farm occurs between the ages of fifty and sixty-four, suggesting that retirement from the labour force to home duties, noted by Massiah, is accompanied by renewed interest in, or need for, cultivating the land. In Nevis, where families are larger, women are most dependent on farm sales below the age of thirty-five, when their children are generally too young to be easily left for long periods with relatives. It has been noted in studies of St Lucian small farming families that younger women give priority to childcare and household duties until their eldest daughter is old enough to take over these chores and thus enable the mother to look for economic opportunities outside the home (Jean 1986).

Figures 6.3 and 6.4 show gender differences in the age at which people farm in Nevis and Barbados. In Nevis, the bimodal distribution indicates that women are most likely to farm when supporting children, or grandchildren as they get older, when many retire from farming and depend on remittances from children usually working overseas. Nevisian men, conversely, spend much of their working life overseas and when

Table 6.6 Sources of income in small-farm households, by gender

Age	Male				Female			
	Own farm	Remittances	Pension	Off-farm work (percentage)	Own farm	Remittances	Pension	Off-farm work
		N=87				N=39		
BARBADOS								
15–34	43	0	0	57	29	14	0	57
35–49	50	0	0	50	33	13	0	54
50–64	33	0	5	62	53	34	0	13
65+	48	1	35	16	39	9	52	0
		N=64				N=34		
NEVIS								
15–34	9	37	0	54	68	32	0	0
35–49	63	4	0	33	47	47	0	6
50–64	56	10	0	34	48	50	0	2
65+	47	29	8	16	52	42	0	6

Sources: Barbados – fieldwork 1987; Nevis – fieldwork 1979.

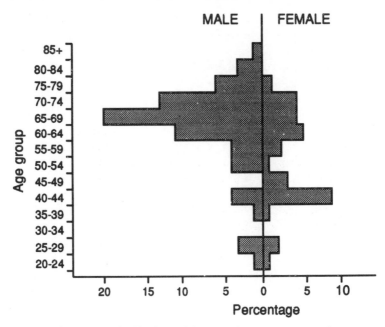

Figure 6.3 Distribution of farmers by age, Nevis, 1979

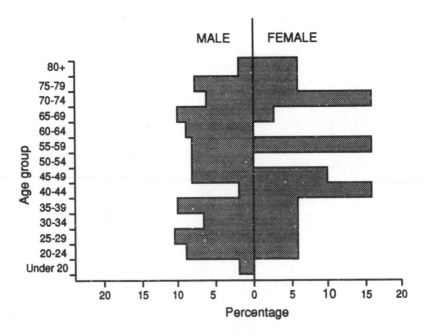

Figure 6.4 Distribution of farmers by age, Barbados, 1987

they retire return to Nevis to farm their small plots of land. In Barbados young men are moving into farming more than young women, and women are more likely to farm after the age of fifty, when farm sales become a major source of income.

Income

In Barbados, remittances are much less important than among the highly migratory population of Nevis, but in both islands they contribute more to women farm operators' incomes than to men's. In both Nevis and Barbados women's dependence on remittances is greatest in the post-reproductive period between the ages of fifty and sixty-four when their children are earning adults. Male and female farmers in Barbados have long been more likely to hold full-time jobs than is the case in Nevis, a difference that is reflected in the importance of pension income in Barbados for those over sixty-five. In Nevis the farm is the major source of income for elderly farmers of both sexes, but in Barbados this is only true for men.

Hours worked

In addition to examining income, we can analyse the changing relative importance of farm and non-farm work through the life course in terms of hours worked (Table 6.7). Because of the small sample size in Barbados, great reliance should not be put on these data, but when considered in tandem with the other information, Table 6.7 does add to the broad understanding of the situation. In both islands, work hours recorded for men are greater than those for women, but most women would spend an additional three to five hours a day on childcare, meal preparation and housework, which are not recorded in Table 6.7. Less time is spent on household duties in Barbados, because of smaller families and the much wider availability of piped water and electricity in homes there. It is believed that this accounts for the difference between the average thirty-five hours a week spent in farm work by women in Nevis and the forty-six hours of Barbadian women. It also contributes to the gender differences in the two islands in hours worked: men in Barbados work only four hours more than women, but in Nevis male work hours exceed women's by an average of twelve hours per week.

Peak farm work hours for both men and women in Nevis occur between the ages of twenty and forty-nine, while in Barbados most farmwork is done by men and women aged between fifty and sixty-five, suggesting that in Barbados there is some element of substitution of farm for non-farm work at this age. In Nevis, women put in most hours

Table 6.7 Seasonal differences in weekly hours worked by male and female operators of small farms in Barbados and Nevis

BARBADOS

	Male N=34				Female N=12			
	Farm work				Farm work			
Age	Peak	Low	Off-farm work	Peak total	Peak	Low	Off-farm work	Peak total
20–34	25	25	34	59	10	8	45	55
35–49	10	6	45	55	17	13	26	43
50–64	31	18	32	63	51	33	16	67
65+	22	16	4	26	17	14	0	17

NEVIS

	Male N=71					Female N=28				
	Farm work		Off-farm work	Marketing	Peak total	Farm work		Off-farm work	Marketing	Peak total
Age	Peak	Low				Peak	Low			
20–34	46	39	4.7	0.3	51	31	19	3	0.6	34.6
35–49	54	43	5.8	0.8	60.6	33	20	3.3	5.1	41.4
50–64	37	26	18.4	2.4	57.8	25	16	1.3	4.6	30.9
65+	37	31	3.3	1.6	41.9	27	22	1.8	4.9	30.7

Sources: Barbados – fieldwork 1987; Nevis – fieldwork 1979.

in off-farm jobs between the ages of thirty-five and thirty-nine, when children are old enough to be left but are still economically dependent. Men farmers spend more time in off-farm jobs at a later stage in life, at age fifty to sixty-five, perhaps reflecting the typical Nevisian male life course in which men migrate in their earlier years, returning to retire on the family farm. Gender differences in off-farm work in Barbados are indicative of the type of work available for women: the market for part-time jobs is predominantly in tourism, which tends to employ mostly young women. Age does not affect male job opportunities so markedly.

The use of female labour as a reserve workforce to cope with peak demand is shown by the much greater seasonal decline in hours worked on the farm for women than for men: low season hours for women in Barbados are 70 per cent of peak, compared to 77 per cent of peak for men, and in Nevis low season hours for women are 66 per cent of peak versus 81 per cent for men. In Nevis both men and women put in the longest hours at work on the farm between the ages of thirty-five and forty-nine, but in Barbados peak farm work hours occur between the ages of fifty and sixty-four. It is at this age that women's hours exceed male hours. This difference is supported by census data which suggest that women retire earlier than men from non-farm employment in Barbados.

TYPE OF JOB AND LIFE COURSE

The type of job done also changes with life stage and age cohort (Table 6.3). Certain jobs, such as clerical, sales and services and production work tend to be dominated by young women, while other jobs such as those defined as administrative and managerial demand years of experience and/or training. As in many developing countries, world market factories employ mostly young women, and Kelly's study of the workforce of St Lucia's export-oriented electronics factories showed that 78 per cent of the workers were women between the ages of sixteen and twenty-seven (Kelly 1987: 8).

The data from Montserrat should not be thought to indicate that women move from factory work when young to agricultural employment when old. Rather, the table illustrates an economy in transition, with the jobs requiring technical skills and training going to better educated, younger women while the older, less well-educated women are left in the traditional service and agricultural jobs. It should be noted, however, that those jobs with the highest proportions of elderly women are in the farming and sales, mainly itinerant trading, sectors where many years of experience are an advantage. A study of female job mobility found that most young women taking factory jobs had previously been unemployed (Skelton 1989).

Marketing is a traditional occupation of older rural women. Spence (1964) in her study of hawkers in Barbados showed that women spend a long apprenticeship, usually working with their hawker mothers. They tend to become hawkers when their children are old enough to allow them adequate time away from home and when the woman herself has gained enough experience and status to be able to negotiate successfully with suppliers and customers. Often the hawker-mother will allow her daughter to inherit her position and will herself either retire to take over household duties or move to a different and usually less arduous type of trade. Either way she will not compete directly with her daughter and will hand over her goodwill with suppliers and customers to her successor.

More recently a study of beach vendors in Barbados (Alcraft 1987) revealed very marked gender differences in age and product sold. Over half the male vendors were between the ages of twenty-one and twenty-five and sold mainly jewellery and aloe to soothe sunburn. Female vendors were predominantly over forty years of age and had a monopoly of sales of clothing and fruit, items traditionally sold by female hawkers. Of the ten women who sold clothes, all had learned to make them from their mothers when they were 'little'. Men saw beach vending as a short-term occupation in a pleasant environment; those surveyed had an average of 6.1 years in the job. Women, by comparison, averaged 12.3 years of experience as beach vendors. They had mostly come into it through an introduction by their mother and continued to vend because of lack of alternative opportunities. Nevertheless, they appreciated the flexibility of the work hours. As one said, 'I used to get babies fast, so it don't make no sense to get ordinary work. Here you come and go when you please' (Alcraft 1987: 23). One-third of the women worked seven days a week, however, and a further 30 per cent worked four days a week. Table 6.7 shows that marketing is more common among women than men in Nevis and that it is an occupation of older women. One woman farm operator aged seventy-one still took vegetables from Nevis to the market in St Kitts twice each week.

Berleant-Schiller and Maurer (1992) demonstrate that women in Barbuda become more publicly visible with age. Their social worlds merge with those of men, after being largely separate during their reproductive years. Older women often run rum shops and bars and also assume local political roles, which they have done most markedly in Dominica. Within households they free younger women for paid employment by taking on the care of grandchildren and are also responsible for the preservation of knowledge of traditional dishes and customs. Brana-Shute (1992), in a study of women's political roles in Surinam also found that women's status and influence increased with age.

CONCLUSION

Reflecting the history of the region and its prevalence of female-headed households, by global standards high proportions of Caribbean women engage in work for pay across much of the life course. In general, rates of participation drop only after the child-bearing years, rather than during them, replicating women's needs to support themselves and their children. At every stage of life, the women's work experience differs from that of men of their peer group. Additionally, though they retire from the workforce at younger ages than men, they often continue in unpaid labour, assisting adult daughters with home duties or expanding their work on small-scale farms to help feed grandchildren. As a consequence, unlike the situation in much of the industrialized world, we find fewer households of single old women in the Caribbean than of single old men. The long-pervasive pattern of migration from the Caribbean also shapes the work lives of older women and their well-being. If adult children are absent, the older woman may find herself caring for grandchildren sent or left at home, while she may benefit from remittance income sent from abroad. Other women leave the workforce and emigrate to join adult children.

Within these general patterns, however, my analysis has revealed differences among the islands in the courses of women's work lives. These reflect the nature of economic development, levels of education and fertility patterns. In smaller and less developed islands like St Vincent and Nevis, where educational levels are still relatively low and families are large, women are likely to farm both at younger ages when they are raising children and in their later years when caring for grandchildren, working off the farm for pay in their middle years. By contrast, in the more developed island of Barbados, where economic opportunities have widened, educational provision expanded and fertility declined, younger women are likely to hold off-farm jobs, with farmwork remaining the domain of older women.

At this point, we cannot predict whether women now in their youth or mid-life will follow their mother's patterns in old age. The availability of some pension benefits may change the need to engage in subsistence production in old age, and the younger cohorts' lack of experience in rural work may also cause them to seek other alternatives in old age. Changing opportunities for emigration, as traditional host countries alter immigration policies, may also affect the roles of older women. As we seek to interpret women's work across the life course in the Caribbean, therefore, we need to see the region both in the context of its history and of its continuing development and relation to the outside world, examining the ways changing circumstances shape the opportunities, capacities and needs of each generation of women.

7

GENDER AND THE LIFE COURSE ON THE FRONTIERS OF SETTLEMENT IN COLOMBIA

Janet G. Townsend

Migration to a new context often means a change in work patterns for women. For example, married women who move from rural to urban areas or who migrate internationally frequently enter the paid labour force in an effort to help the family survive in its new setting. Single women migrants work to support themselves and may also send resources home to their families. In Latin America, migration has been best documented for girls and women moving to urban areas (see, for example, Young 1982). But another important form of migration in a number of Latin American countries involves movement to pioneer zones in the forests on the frontiers of settlement. In contrast to urban migration, rural-to-rural migration to new land settlements often means that women lose rights in land, lose direct income, face increasing workloads, particularly in reproduction as education and health services are poorer, and suffer painful social isolation. Such changes have been documented in new land settlements in Bolivia, Brazil, Kenya, Malaysia, Nigeria, Sri Lanka and Zimbabwe (Townsend forthcoming b); generally they occur where the settlements include unusually high proportions of men.

Why do these changes occur? Do they only affect mature married women with children? What happens to the younger women and girls in pioneer families as they grow up? What happens to pioneer women as they age and the settlement becomes more established? How does the specific context of the pioneer area shape women's experience within it?

In an attempt to gain insight into these questions, I have conducted research in three contrasting small pioneer areas in Colombia. I shall first describe the nature of these areas, then examine the gender division of productive and reproductive work within them, looking at males and females of different ages via a cross-sectional analysis. I shall then provide case studies of selected households, showing how experiences have varied over the life course. These case studies provide an opportunity to explore differences between an individual's life path and the collective patterns revealed by the cross-sectional material. Through these analyses, I hope to reveal some of the ways in which women's lives are shaped by both their context and their life course, and to examine the choices people make as they deal with the constraints and opportunities of living in these pioneer settlements.

GENDER ROLES AND TIME IN THE FOREST LANDS OF LATIN AMERICA

People who seek to farm in the forests of Latin America experience multiple transitions – in their own biological and social lives, and in the contexts in which their lives are placed. First, the farming systems themselves are being transformed as the land is brought into cultivation and also as the economy is integrated into the market system. Frequently, the first stage of the settlement after land-clearing involves cultivation of crops, but for many farmers, ranchers and speculators in land, the longer-term objective is to engage in cattle rearing. Second, the pioneer families age chronologically. Children grow up and are able to take on new tasks or are directed towards education. As a woman bears children over the years, both her household burdens and her possibilities for assistance change, though these will vary depending on the number, spacing and ages of the children in the household at any one time. As children leave the household, as people become aged and physically less able, and as old people die, responsibilities and options continue to change. To understand an individual life course in such pioneer settlements, it is important to think about how these two sets of transitions coalesce. For a woman who comes to the pioneer community as a young girl, the experience will be different from that of one who comes as a married woman who already has children and an array of household responsibilities.

Other studies in South America have suggested four different sequences of changes in gender roles in pioneer settlements, each with a different dominant process (Townsend forthcoming a). One pattern, revealed in eastern Amazonia, involves capitalist penetration of the farming system. At first, women and men pioneers work together to make family farms out of the forest. After some years (from five to thirty, or even longer, depending on location), these farms are displaced by ranches and the rural women withdraw from work on the land to become housewives. A second pattern, identified in eastern Colombia, shows changes that reflect both the passage of historical time and family time. As children grow up and are able to undertake more agricultural work, women who initially worked on the farm withdraw from heavy field labour. Profits from coca production also permit male labourers to be hired. But women remain active in lighter agricultural work and in ranching. In eastern, lowland Bolivia, a third pattern has been recorded. In their previous homes in the cool highlands women had abundant opportunities to generate and control income as traders, as artisans and as producers of small livestock on communal land. In the pioneer colonies, these opportunities disappear, resulting in new gender divisions of labour under which women work on their husband's land. The remaining pattern noted involves a sequence of migrations – settlers move to the pioneer area, but work patterns emerge which create an apparent surplus of female labour and a demand for male labour in the pioneer area. Under these conditions, a new migration occurs, whereby young women leave the pioneer area.[1] It is this fourth experience that I will examine in this chapter.

SETTLEMENTS IN THE MIDDLE MAGDALENA VALLEY

My research has been carried out in three areas in the middle Magdalena valley of Colombia. Over the last thirty years the forests in this hot, humid region have largely been cleared for ranching. Within this common setting, the three areas are quite different from one another in their demographic structure, economic organization, integration into the national economy, prosperity and accessibility. Despite these differences, the gender divisions of labour which I identified in the three settings are remarkably similar – women are confined very strongly to reproductive roles. In all three areas, settlement is highly dispersed: households live on their farms or ranches.

In San Lucas, which consists of several small villages and hamlets, the development of settlement and the pace of change have been relatively slow. Although the pioneers began to clear the forest nearly forty years ago, no roads have been built; shifting cultivation is still the practice on the crop farms, and ranches have not yet taken over. Living conditions are poor and services of any kind are few and far between. Many farmers are isolated in the forest. Few families are landless, although many men work as wage labourers. Industrial inputs to farming are few: the main tool is the machete, and farming and transport depend on human and animal power. In the slack season, men migrate out to work as casual and temporary farm labourers in other areas all over western Colombia. Households are large and fertility is high. Some 40 per cent of the population is under thirteen years of age and only 9 per cent is over fifty (Table 7.1). Hands seem to be at a premium; yet women are almost excluded from work on the land and many girls leave the area at an early age. Indeed, above the age of about ten, females are in a minority in San Lucas. The sex ratio among adults (fifteen years of age

140

Table 7.1 Population structures by gender and age in each survey community

Locality	Children		Adults		Over 65	Total
	under 7	7–12	13–50	51–65		
San Lucas	22	19	51	6	3	100%
f:m	11:11	9:10	18:33	2:4	1:2	40:60
	(1:1)	(0.90:1)	(0.55:1)	(0.50:1)	(0.50:1)	(0.67:1)
El Distrito	15	14	52	7	13	100%
f:m	7:8	5:9	24:28	2:5	2:11	40:60
	(0.88:1)	(0.86:1)	(0.88:1)	(0.40:1)	(0.18:1)	(0.67:1)
La Payoa	28	19	48	3	1	100%
f:m	15:13	10:9	22:26	1:2	0:1	48:52
	(1.2:1)	(1.1:1)	(0.85:1)	(0.50:1)	(0:1)	(0.92:1)

and over) in San Lucas in the seventy-five households sampled in 1984 was 189 men to 100 women; a third of the total population were adult males. Although San Lucas produces some surplus maize and meat for sale in urban markets, nationally it is primarily a source of reserve labour – seasonal male workers for commercial agriculture, and female domestic servants for the cities.

When I surveyed the other two communities, they were no longer at the frontier. Both El Distrito and La Payoa were created as projects for the establishment of family farms in places where ranching had already replaced the forest and led to conditions of poor employment and productivity. El Distrito, only 100 km south-east from San Lucas has undergone quite different historical change since it was first settled thirty years ago, and is now strongly incorporated into the urban-industrial economy. The nearest city, Bucaramanga, is only five hours away by truck. El Distrito is an example of high-cost land reform: in the 1960s, the state bought ranches and drained and subdivided them; later it used a World Bank loan to develop irrigation. Now the district has 211 family-owned farms producing cash crops (meat, milk and rice) with the assistance of seasonal hired labour from surrounding villages. The technological level of operations is much higher than in San Lucas – a third of the families surveyed in 1987 had their own tractor, some had combine harvesters, and many practised aerial spraying of rice. Production is energy-intensive, with little place for permanent family labour, even though non-family labour is needed irregularly. Demographic conditions are also quite different from San Lucas. Completed families are small and households even smaller. Less than a third of the population is under thirteen years of age, but there is a much larger presence of older people, mainly men – 20 per cent of the population were over fifty (Table 7.1). Again men predominated, the adult sex ratio in the thirty households sampled being 153 men to 100 women.

La Payoa, 50 km south of El Distrito is also a new project for landless families from the nearby mountains, developed in 1986 by Servicio Pastoral de San Gil (SEPAS), a Catholic organization. It set up a cooperative of fifty-eight families with the formal aim of implementing a farming system that would be peasant-based, grass-roots-oriented and use small-scale 'appropriate' technology. One intent was to maximize autonomy from the market. The cooperative still has to pay off the purchase price of US$500,000 for the old ranches, however, so that it has little immediate freedom to withdraw from the market. By September 1987, just over one year after the first settlers arrived, many teenage girls were planning to leave, but the labour force had been much expanded by older male relatives brought in from the mountains. I shall later consider possible reasons for this. The adult sex ratio in the forty-six households I surveyed was 136 men per 100 women. Most of the households were youthful: nearly half the population was under thirteen years of age, with only 4 per cent over fifty (Table 7.1).

Given their different patterns of economic organization and relation to the market, it is not surprising that the class structures of the three areas should also differ. If we make a crude division, three classes of household can be identified – an upper class of landowners, merchants and rich peasant households which hire labour; a middle class of peasants able to achieve a livelihood with family land and labour; and a class of poor peasants who sell their labour. Using this classification, we can recognize all classes of households in San Lucas, only rich peasant households in El Distrito and only middle peasants in La Payoa. Hired workers who reside in the households of their employers comprise the poorest group. These include female domestic servants (usually with small children and no partner) and male labourers; sometimes a couple will combine these duties with that of caretaker. Resident male labourers have a more circumscribed role in production than male family members and perform less skilled tasks, often in agriculture rather than with cattle; many in El Distrito are elderly. Class differences among the households make little difference in the types of work women do, though more prosperous women may have the quantity of their work reduced by a domestic servant who does the full range of reproductive tasks. These servants always live in the household of their employer, even if they are married to a male labourer. They do double reproductive work – for their own children if single, for husband and children if married, and for their employer's family.

The class distinctions are reflected to some extent in the educational levels of people in the three communities, but gender and generation are more significant causes of educational difference. Women of all ages have less education than men, but as education reached women later, the differences are more pronounced among the elderly. Older people (men over sixty-five and women over fifty-five) have had no formal

142

schooling; few older men and no older women can read. The recent spread of education has modified the past sharp discrimination between the genders, however, among the younger age groups. Comparing the three areas, we find that San Lucas has the highest levels of illiteracy, with people of both genders, all ages and even all classes who cannot read. In El Distrito and La Payoa, few adults under forty are illiterate, although few have completed primary school. Of the few in all three locations who have completed primary school, all are under thirty-five and only about a third are female (even less in San Lucas). No one has completed secondary school, and no woman over twenty-five has had any. In El Distrito and La Payoa, some people have attended training courses (in farm matters for men, domestic matters and handicrafts for women). Again, more males than females have attended courses. The exceptional case was in a household in El Distrito comprised of a farmer and his *de facto* wife; she had some secondary education and did the farm accounts. His legal wife and children lived away in the city where the children were receiving secondary and higher education.

WORK AND THE LIFE COURSE

Within these pioneer contexts, how do gender and life stage interact to shape people's work lives? The stereotype is that a young couple comes to a forest area, acquires a piece of land and sets out to make a farm. It identifies children as an economic burden when young, but describes them as soon becoming an invaluable source of labour, certainly by the age of thirteen. As the children grow up, marry and have children, an extended family with its associated networks evolves, working in households but exchanging some labour and support. This is the history which many women recounted to Donny Meertens (1988) in eastern Colombia, and it would fit some of the early settlers in the middle Magdalena, but only a few of the families I interviewed. Some of the new arrivals were young couples, families in the early phase of formation, but not all. The labour force is recruited from siblings, labourers and fictive kin, as well as from children. In San Lucas, where the economy is least monetized, half the households have become compound by recruiting extra-'family' labour (such as distant, adopted fictive kin), while in El Distrito, half the households have resident hired labour. It should also be noted that farm families in San Lucas and El Distrito have little continuity because of the high rate of economic failure, which causes many families to move away.

In investigating the life course experience, I have focused on work, rather than other activities, because in these areas, work extends almost across the entire life course, especially in San Lucas, the most labour intensive of the settlements, where those who 'always participate' in tasks ranged from small children to the elderly.

Table 7.2 Female : male ratios by age group in each survey community of those who 'sometimes' or 'always' participate in tasks

| | Children | | Adults | | Over 65 |
	under 7	7–12	13–50	51–65	
SAN LUCAS					
Total sample	60:61	50:55	104:187	9:22	5:11
Tasks					
Workers' meals	0:1	4:1	43:0	5:0	–
Laundry	5:1	30:0	93:21	7:4	3:3
Family meals	2:1	31:11	99:57	9:7	3:4
Feed hens	4:2	21:7	79:37	7:6	2:4
Feed pigs	2:2	11:11	43:48	3:6	2:2
Fetch water	10:10	38:35	75:95	3:9	1:5
Fetch wood	2:6	21:36	33:143	1:14	1:7
Gardening	–	0:4	3:18	1:3	1:4
Milking	0:1	1:2	4:53	0:5	0:4
Sowing	0:1	0:13	5:127	1:16	0:5
Fell trees	–	0:7	3:96	0:12	0:2
Cattle care	0:1	0:11	0:70	0:8	0:2
EL DISTRITO					
Total sample	13:15	9:16	45:52	4:9	4:20
Tasks					
Workers' meals	–	2:1	40:3	4:1	–
Laundry	–	6:1	40:4	4:1	3:4
Family meals	–	3:1	43:2	1:1	3:4
Feed hens	–	2:2	40:5	3:1	3:4
Feed pigs	–	–	8:1	1:0	–
Fetch water	–	2:3	29:5	2:1	1:4
Fetch wood	–	2:4	34:8	2:1	1:4
Gardening	–	–	1:5	1:0	0:1
Milking	–	–	8:33	0:4	1:8
Sowing	–	2:6	1:48	–	0:18
Fell trees	–	–	–	–	–
Cattle care	–	0:1	2:47	0:6	0:19
LA PAYOA					
Total sample	42:37	29:26	60:72	3:6	0:2
Tasks					
Workers' meals	–	4:1	34:3	1:0	–
Laundry	–	10:5	54:5	3:2	–
Family meals	–	13:7	57:4	3:1	–
Feed hens	2:2	4:6	48:9	2:0	–
Feed pigs	–	3:3	33:7	–	–
Fetch water	–	8:10	37:17	3:0	0:1
Fetch wood	–	10:13	53:23	3:1	0:1
Gardening	–	–	1:2	–	–
Milking	–	2:2	33:38	0:2	0:1
Sowing	–	0:5	2:66	0:5	0:2
Fell trees	–	–	–	–	–
Cattle care	–	–	6:56	0:3	–

Note: Tasks are ranked according to their position among adults in San Lucas, from the most female, to the most male.

Work is highly gendered throughout the life course. Reproductive activities are the responsibility of females, productive activities of males, with very few exceptions (such as men building and repairing the house). Girls take up domestic labour and collect water and wood, and are fully occupied in all household tasks from the age of about thirteen. Woman's archetypal role is in the house throughout her life. With very few exceptions, from childhood to old age women participate in cooking for the household, cooking for labourers, cleaning, washing clothes and caring for children, although after the age of fifty most of them are less engaged in heavy tasks such as carrying water and wood, and the intensity of their work in general declines (Table 7.2). At no time do they participate significantly in the generation of income for the household.

Boys also fetch water and wood. Otherwise, the contribution of boys and young men to domestic tasks is small, and is occasional rather than regular; adult males participate in domestic labour more than boys, but many of them only in emergencies. Boys begin early to contribute to agricultural field tasks and the care of cattle, and are fully involved from their early teens unless they are at school. Again, work is for a lifetime: for most older men, the work they do may be less effective, but no less time-consuming or demanding than earlier in their lives. Participation by old men is reduced more than that of old women, but older men take on a greater share of domestic tasks than adult men under fifty (Table 7.2).

When I plotted activities by gender and age, I found that the course of a working life revealed a regular collective pattern: dependency (until about seven years of age), apprenticeship (roughly from seven to twelve years), adulthood (thirteen to sixty-five) and old age (over sixty-five). Children under five are rather vulnerable consumers, who make work for the women and older children who strive to keep them clean, fed and healthy in the face of both nutritional difficulties and parasitic and respiratory diseases. Indeed, the physical condition of the children differs considerably among the communities, though the class of household and the position of the child in the family also influences their condition. In El Distrito, for example, collective services such as vaccination help to improve young children's well-being, whereas such services are effectively absent in San Lucas. Further, in San Lucas, child malnutrition is widespread, though none of the cases of second- and third-degree malnutrition in the survey was the child of the senior male in the household or of his sons. Despite the expressed preference for boys, there is no gender difference in nutritional status (Townsend and Wilson de Acosta 1987).

Only in poor, labour-intensive San Lucas do small children sometimes participate in household tasks, particularly fetching water (Table 7.2). Children from five to seven can do more for themselves, and some share in household chores. Their place is very much in the house, however, as forest and paddocks (with 'savage' cows) are considered too dangerous for them, and the garden may not be fenced. To the visitor, households appear centred upon small children, who are the focus of attention, even of men, when they are in the house.

Children from seven to twelve may make a significant contribution in the work of the household, especially in poor families and in San Lucas where human labour is the main source of energy. This work ties girls to the house but for boys may mean going some distance to work in the fields or paddocks. School, if they attend, greatly reduces children's contribution to work and broadens their activity space (especially for girls) because even attending the village school may mean being away from home for much of the year; a few children go to school in the city, where they stay with relatives. Many girls begin to leave school after about ten, particularly from San Lucas. They can rarely add to household income, and as they become more demanding as consumers, their reproductive contributions to their households of origin are not worth the cost of maintaining them. An elderly dependant doing similar work and incurring the same costs would be accepted as a responsibility, but adolescent girls, on the whole, are not. This situation often provokes tension within the household and, if young women do not find a partner, they may leave for the city. It is at this point in the life course that gender distinctions become heightened and lead to different patterns in males' and females' life choices. Again, it is in San Lucas that children contribute most to household tasks and even work with cattle or felling trees; in El Distrito, they do least. Their participation is in inverse ratio to the wealth of the area, and also to the availability of schools.

Adults (thirteen to sixty-five) face a time of stress and disturbance, which may recur through their adult working lives in forming and dissolving partnerships, in procreation and/or in migration. Women's and men's tasks vary little, as we have seen, through the adult years. The intensity of work, however, may depend much on their position in the household, even in San Lucas. In El Distrito, prosperity brings labour-saving devices into the home as well as the field; a few families have generators, blenders, water pumps and piped water, as well as tractors and combine harvesters. In all areas, family stage influences women's lives in particular, depending on how many children they bear and on their sexes and ages.

Very few people remain single, and few remain childless after their early twenties. A woman's life in the middle Magdalena valley, probably more than elsewhere in Colombia, is seen as being for biological reproduction. Completed families are large – many women have borne more than ten children; one respondent had had eighteen pregnancies.

146

Paramilitary 'death squads' claim to eliminate homosexuals ('Death to communists and homosexuals!'), and certainly sexuality is discussed only in heterosexual terms. A woman's sexuality is for the establishment and maintenance of the conjugal relationship upon which she and her children depend for their economic survival. Marriage is usual in La Payoa but rare in San Lucas, where partners may change several times during the life course. For women this may include excursions into prostitution. Patterns in El Distrito are between these extremes. A woman with children who has lost her partner and cannot find another may work in other households as a domestic servant. The children of these women tend to be the most disadvantaged in the area. A man's life is defined in terms of his work, income and number of children; a woman's in terms of her achievements in the sphere of social reproduction and, on occasion, in terms of her chastity.

Within the lives of adults, migration is an important component. Men often had circular migration patterns, whereas women who migrated returned less often. Many men cannot afford to maintain a wife or partner, so many girls and women leave the communities for the cities. This phenomenon is particularly evident in San Lucas. In later years, these migrant women may bring their children back and leave them to grow up on the farm, again a pattern particularly noted in San Lucas. Movements to kin are important and many families maintain very impressive networks. It is normal for adults in the valley to have siblings and/or children all over the country, engaged in activities ranging from domestic service and casual labour to university education.

The experience of old age (over sixty-five) depends greatly on a person's cumulative life history and on the position achieved in the household and community. The differences between the lives of the senior couple in an extended family household and others who may live with a household, such as an elderly uncle or employee, are much greater than those between owners and dependants at an earlier age. It is among older people that status overwhelmingly overrides chronological age, for this may be a time of great authority, or dependence and humiliation. Young children in the household show no respect to elderly male employees. Isolation in old age is rare; if an elderly person has any relatives in the area, s/he will live with them (but see the case of Alirio, below). The plight of elderly employees is grim, even if they live in a household. (One option for elderly males is to be a low-wage live-in labourer in prosperous households, particularly in El Distrito (Tables 7.1, 7.2).) While lifelong agricultural labourers may be entitled to some state pension benefits, there are no pensions for women who have spent their lives in domestic service, nor for women or men who have been self-employed. For the elderly, differences between the three areas are of degree rather than kind within the patterns just described: again, the differences in wealth and in the provision of collective services between the three areas are significant, shaping the variations in experience of the elderly.

In all three areas, then, the context of a working life is strongly set by the productive economy of the farm, and of the area, and by the accessibility of collective services. For the individual, date of birth is very significant, given the scale of secular change, as are transitions to new positions in the household, but gender remains the overall determinant of the work life.

GENDER ROLES AND THE LIFE COURSE

The patterns I have described reveal that despite the differences in economic activities and the class structure of households among the three areas, the life courses of the women are remarkably similar – those who remain on the frontier are excluded from income-generating work, and many women are effectively expelled from the region. While relative exclusion of women from farm work is common in rural Latin America, the middle Magdalena valley examples represent an extreme case.

The patterns of work and migration are interrelated. One reason why reproductive work commands so much of the women's time is that there are so few women, and, given that reproduction is considered women's work, they have time for little else. In eastern Colombia, the preponderance of males is a function of the large proportion of single male labourers, rather than sexual imbalance within farm families (Meertens 1988). In the middle Magdalena, however, the shortage of females is within the family. The male predominance in the middle Magdalena is produced by migration: among the older population, more males than females migrate in (Table 7.1); among the younger population, more females than males migrate out.

The origins of these patterns are slightly different among the communities, reflecting their historical origins and contexts. In San Lucas, the remarkable predominance of men appears to have been created by the arrival of larger numbers of men than women and sustained by the selective outmigration of girls (Townsend and Wilson de Acosta 1987). I did not interview women and girls who had left, but it would be almost impossible for a child to leave San Lucas without family support, given the difficulty and cost of the journey out by trail, river and road, so it must be concluded that they left with the approval of their families. The male population in San Lucas is variable because circular seasonal migration is important. Selective migration also creates the masculine sex ratios in El Distrito, starting with young girls being sent away. The greater prosperity of El Distrito allows this, with migration usually being

Plate 7.1 El Distrito: a girl of fourteen about to leave the
region, forming part of female out-migration

to attend school. In general, they do not return; older girls and women
also leave, as do a smaller number of boys and men, usually on a more
permanent basis than in San Lucas. In La Payoa, which was founded
only in 1986, the male preponderance mainly reflects the composition
of the new adult settlers.

Gender roles in income-generating work on the farms almost exclude
women and girls (Tables 7.2, 7.3). This is true even in San Lucas, the
low energy input, labour-intensive case. People explain this by the
'savage cattle': the lowland, Cebu cattle are seen as wild, dangerous and

unmanageable by women. Dangerous cows also present a spatial constraint for women and children – their presence defines whole spaces as threatening, as paths cross paddocks. Certainly injuries are commonly received from cows, not bulls, but this is a function of cattle management; a ranch cow with a calf will have had little handling and can be a real danger. Still, families with only one or two cows may make no effort to tame them here, although women milk and handle the same breeds in eastern Colombia and some women do in La Payoa (Tables 7.2, 7.3). Women's almost complete exclusion from cultivation and even gardening in San Lucas is unexpected (Tables 7.2, 7.3). To me, it is also unexplained, although the very small number of females clearly plays a part, as does the high proportion of pioneers who came from the northern coastal plains where women's participation in cultivation is relatively low. Women's main direct contribution to household income, as Carmen Diana Deere and Magdalena Leon de Leal (1983) found elsewhere in Colombia and in Peru, is in cooking for the labourers, who are partially paid in meals. Men are more active in reproductive tasks than is usual in rural Colombia, particularly in San Lucas, where the sex ratio among adults in the 13–50 year age group shows the most male dominance. Perhaps because women are in very short supply there, men contribute significantly to reproductive tasks which are held in high regard. In La Payoa, where the sex ratio in the same age group is closest to equal, such tasks are more exclusively performed by women.

Excluded from agricultural work, and bearing heavy household responsibilities, adult women in these areas lead very isolated lives, perhaps seeing only other members of the household for months on end. Both their activity spaces and social networks are very constrained. About the only opportunity to leave the house is to do the laundry,

Table 7.3 Female : male ratios of adults (aged over twelve) in each survey community who 'always participate' in tasks

Tasks	San Lucas	El Distrito	La Payoa
Workers' meals	43:0	7:1	16:1
Laundry	10:1	8:1	18:1
Family meals	6:1	8:1	18:1
Feed hens	3:1	8:1	11:1
Fetch water	1:1	4:1	3:1
Feed pigs	1:1	7:1	15:1
Fetch wood	1:6	4:1	3:1
Gardening	1:6	1:3	1:2
Milking	1:25	1:7	1:1
Agriculture	1:60	1:52	1:70
Cattle work	0:65	1:28	1:13
Total population, females to males aged over twelve	1:1.9	1:1.5	1:1.3

necessitated by the rarity of piped water in each case, even in El Distrito. Some respondents in San Lucas had had to deliver their own babies because they knew no one to ask for help. Only in La Payoa do women really know other women outside the family, because the cooperative has brought them together. Shopping may provide an opportunity for outings in El Distrito or La Payoa, but in San Lucas shopping is a male activity in farm families, as the village is hours or a day's walk away. Among the only adult women who escape this isolation are the prosperous, who accompany their children to the towns and villages where they attend school, setting up homes there during the term time. Other women who work as prostitutes in nearby villages or towns, or who leave for the cities, may in part be fleeing this isolation.

For those who come to the middle Magdalena as young girls, migration to the frontier is a temporary phase in their lives; many leave for schooling or domestic employment and make new lives elsewhere. For women who come to the pioneer zone as adults from settled agricultural areas in other parts of the country, the new life represents a radical transition. In their home areas they lived among a network of female relatives and often relied on them heavily for social satisfaction and personal support – relying particularly on their mothers. In all three areas, they report that in their home viilages they felt safe to move freely, whereas now they live among strangers, as well as being heavily constrained by their work roles.

My research yielded only limited information on gender relations which are so central to life transitions. In formal terms, men have great power, though we shall see below that women's informal power may also be considerable. Alfredo Molano (pers. comm.) has suggested that, by Colombian standards, a woman's position in land settlement areas is unusually strong, for her work is critical to the survival of the farm enterprise and her bargaining position is therefore unusually good. Certainly the abandonment of farms in San Lucas is described as a result of the departure of women. Formally, however, it is pioneer men who obtain title to land. A woman may inherit land, but usually depends on a new partner, a grown son or a hired male manager, to work or run it.

Age is important in power, as are economic status and the role as head of household. Women's informal power appears to increase greatly with age, perhaps as a return on social investment in the family, perhaps with a continuing development of social skills. Still, many women must confront violence from men, especially when men have been drinking. Though today this is less condoned verbally than in the past, it is not clear whether or how women's persistence or departure from the area is connected to their marital relationships. My general conclusion from an examination of the experience of adult migrant women is that it represents an extreme case of what Maria Mies (1986) has termed 'housewifisation', a decline in participation in paid productive work that creates an ensuing dependence on others for income. Total work may actually increase, but becomes invisible, and direct access to income is lost. It is a pattern that has been identified by Chambers (1969) in other areas of pioneer settlement where women have lost direct income and land rights, suffered an increase in workload, have few services to support them and face isolation, compounded by the transition from extended to nuclear family life. The degree of relegation to the house in the Magdalena case is especially marked.

INDIVIDUAL LIFE COURSES

Though this cross-sectional analysis reveals two general patterns of migration and work among girls and women of the pioneer colonies, it cannot show how individuals deal with the choices and constraints they face at different points over the life course. Are the choices for most simply between migration to the city (and domestic service) or a submissive isolated life with a husband in the valley? What options are open to women who stay? Longitudinal analysis can provide insights to these questions. Here I shall present brief case studies of selected households to illustrate diversity within the general patterns.

Rosa and Alirio

Even for men with reproductive success, outcomes are not standard. Alirio is seventy-six and his wife Rosa is seventy-two. Born in the mountains, both reached El Distrito after years of migration, tenant farming and sharecropping. With five children and over twenty grand-children, Rosa and Alirio have achieved considerable reproductive success. Their offspring have been upwardly mobile; at the time of my study, each lived in a different town. Rosa now lives with one of her grown children in the city 'because of the climate', while Alirio lives alone and in some poverty on the farm. Alirio was a client of the state agrarian bank for thirty years; but credit is not available to farmers over sixty, and he now grazes other people's cattle on land which he rents. An elderly labourer lives with him and does the farm work. None of his family visits, although they do, irregularly, send money.

Etilda and Justo

In contrast, Etilda, fifty-six, has made Justo, sixty-five, a successful patriarch in San Lucas. Their first two children were born on her father's farm, only a few days' travel from San Lucas. In 1967 they became settlers in the forest, bringing almost no resources save their small boys and Justo's two small boys by a previous union. They had another six boys and now Etilda's eight sons all live and work on the farm, as do a granddaughter and nephew and Justo's eldest son. They had two daughters, one of whom, now married, lives nearby with her family, as does a married son. With this family labour they have made a farm of 300 hectares, to which they have title, and they own thirty head of cattle. On the farm, Etilda and the seven-year-old granddaughter do almost all the domestic labour; the granddaughter is Justo's daughter's child, brought 'from the city' by her father and abandoned on the farm. In this household, reproductive activities are referred to with considerable respect, and Etilda with love and pride: the farm is considered as much her creation as his, even though she has never participated in farm work.

Carmen and Pablo

At La Payoa, Carmen and Pablo are both 'about fifty'. He was a union activist among the sisal workers in Onzaga, where they came from eighteen months previously. They are founder-members and leaders of the cooperative, and both participate vigorously in cooperative meetings. Carmen noted that women who have moved to the cooperative have been deprived of direct income. She is eager to identify income-generating tasks for women. She raises pigs and sells clothes on commission, the profits going into the household budget. In Onzaga, says Carmen, 'women don't really work outside' but there she made handicrafts from sisal at home and paid for the children's clothing and often for their food with her earnings. She also had a kitchen garden, kept pigs and raised poultry. In La Payoa she tried to keep pigs, but found the new environment difficult. She helps to milk the cows, as does one of her daughters (as we have seen, far more effort is made in La Payoa to domesticate cows (Table 7.2)). Five of Carmen's and Pablo's eleven children live with them; one lives in the city; and five live elsewhere in La Payoa as members of the cooperative. They have eight grandchildren. Carmen's economic activity is no measure of her power. With the fifty-eight members of the cooperative including not only her own household but also those of five of her children, Carmen is establishing herself as one of the most powerful people in the community.

Elena

Finally, the situation of a young single woman may be illustrated by Elena, the secretary of the cooperative at La Payoa. Her story reveals how women's choices become restricted. Elena is twenty-six and has come from the highlands to do the cooperative accounts for a wage and as an act of social commitment. She is better educated than any of the members. Elena's position is marginal, excluded. Unlike the cooperative members, but like the three young women employed by the cooperative as primary teachers, she works short hours and is extremely bored. The cooperative is desperately short of skills: not only skills with cattle, tractors and local agriculture, but in accounting, clerical work and dealing with the bank and the bureaucracy. Elena and the teachers are a considerable potential resource for education and organization. The cooperative manager is a peasant leader with entrepreneurial skills and some knowledge of farming and cattle. He works extremely long hours at these activities, and at motivating the cooperative members to ever greater herculean efforts, but he (like the cooperative council) perceives young women as transient and unprofitable. They are not regarded as potential partners, let alone leaders, but as people who will complete their service and leave. Elena has no place in La Payoa; neither have the more educated daughters of cooperative members. All these girls plan to leave because there are, indeed, no opportunities.

CONCLUSION

The study of work sheds some light on the life course, but for decisions on policy we need to study other transitions in people's lives. Quite simply, we know little of women's practical and strategic needs (Moser 1987) in these areas, and even less of women's strengths. We know that they have little access to land title, to training or to credit, and it will be of fundamental importance that this be changed. But we know too much of women's weaknesses, such as economic exclusion and isolation, and far too little of the bases of women's power: Etilda and Carmen have succeeded through the existing social institutions, not in spite of them. External intervention could easily impoverish women's choices more than it enriches them. These women are active in constructing their own lives and should be participants in any planned change.

As these women and men construct their lives, we see that they are actors in a whole web of opportunities and constraints. The intersections of gender, age, marital status, class, place and moment in time seem to determine many of their opportunities to choose. For instance, a woman who arrived in the area with her husband to try to make a farm when there was still forest available seems, if she decided to stay, to have had little option other than to become a housewife, isolated with her family. If they were successful, or if they came later as prosperous purchasers, she will have had the choice of staying on the farm or of maintaining a second, urban residence with the children at school, and of perhaps making a new life with them in the city in later years. For poor couples arriving now, the openings are few and very limited, with the chances of upward mobility minuscule. A poor single woman arriving must find a partner or choose between domestic service and prostitution; the likelihood of her children being well enough fed to fulfil their basic genetic potential is low.

Girls who arrived as teenagers will have had few opportunities and many will have left, particularly as they may well have been set apart by a higher educational level than that locally available or valued. The opportunities of children born here are strongly determined by gender and class: an affluent girl may escape through education, a poor one only through domestic service or prostitution, having taken on an 'adult' workload by the age of thirteen.

Choices made by parents may thus affect the choices open to children; conversely, choices made by children may affect the choices open to parents in old age, which in San Lucas may begin at fifty. Retirement to the town or city as dependants of children who have left with even a little success is perceived as an attractive option; living conditions in the area are seen as taxing for adults and undesirable for the old. Daughters are considered to be more willing to accept responsibility for old or sick parents, but are less likely to have the appropriate material conditions than sons. Sons, as we have seen, tend less than daughters to leave the area permanently, but some do, particularly if they come from an affluent family and are therefore educated.

It is not only the choices made by children but their very gender which may have powerful effects on the course of their parents' lives. Justo's farm is Etilda's creation in part because she bore him eight sons and apparently only two daughters: the farm was built out of male family labour from the selva. Ana, also in San Lucas, is 'cursed' with daughters and a predominantly female household: two of the three grandchildren living in the household have third-degree malnutrition, while the third is still being breast-fed but will probably also become malnourished. Nevertheless, household composition leaves choices open, and does not determine behaviour. Of 104 adult women in the San Lucas sample, three 'always participate' in agricultural field tasks. All are poor, but none from Ana's household. However, very few households in San Lucas have more women than men; and those which do are all very poor.

Such disparate phenomena as accidents of birth, the date of arrival and the choices of their grandchildren may impinge on and even reshape the life courses of the people of the middle Magdalena valley. Even the success of the women of La Payoa in avoiding the usual isolation may be traced to an event in time, rather than to the ideology of the cooperative. The families of the cooperative arrived together and lived together, crowded into the big buildings of the old ranches while they built their own houses: they have a history of shared hardship and shared physical closeness in common.

155

8

OLD TIES

Women, work and ageing in a coal-mining community in West Virginia

Patricia Sachs

INTRODUCTION

Every community comes into being at a certain moment in history. The nature of the community, the design and structure of its houses and the occupations of its inhabitants all reflect the era in which the community was created and the changes it has undergone.

Deckers Creek was built at the turn of the century. Like other such communities throughout Appalachia, the mountainous region in the central eastern United States, it was originally a 'coal camp' composed of poorly-built houses, outhouses and coal sheds. In the late 1970s, when Deckers Creek was home to retired coal-mining families, it had been transformed into two neat rows of houses facing one another along a tree-lined street, each with a trimmed lawn, flower beds and well-tended gardens. The physical transformation of Deckers Creek has masked its humble beginnings through decades of domestic work invested in the houses. This local shift took place at the same time that the region was altered by shifting economic forces.

In this chapter I describe and analyse the work – especially women's work – that has been carried out in Deckers Creek[1] over a period of sixty years. The physical transformation of the community is a result of the labours of women and men whose values relating to work were formed during their early lives in the coal-mines and company towns. The painted houses and trimmed lawns of Deckers Creek are symbolic representations of the value of community ties and support that is played out through domestic labour during old age.

THE COMMUNITY

In the late 1970s Deckers Creek was largely inhabited by older couples, widows and widowers, who had lived there since the days when the Deckers Creek mine was active.[2] At that time there were eighty-seven people who lived in the forty-two houses that composed the community.[3]

Just over half (52 per cent) the residents were elderly (over age sixty-five). While the percentage of older residents does not suggest a dominance of elders, the residence pattern does: 69 per cent of the homes were occupied by them, with nearly one-third (30.9 per cent) occupied by widowed residents. Eight houses (19 per cent) were occupied by middle-aged (forty to sixty-five) couples, and five (11 per cent) belonged to young couples. None of the middle-aged and young people in the community were related to the older residents.[4] Their own children had left the region during the 1950s and 1960s, when coal-mining had so declined that few jobs had been available for them to fill. The new residents of Deckers Creek arrived in the late 1960s and 1970s, as coal began to boom. Jobs outside the mines had increased at the same time. None of the younger residents ever worked in the mines, just as none of the older residents ever held retail or professional jobs. A 'generation gap' existed along three dimensions: age, the lack of a kin relationship and work history.

Although there were nearly as many young as old people in Deckers Creek, their impact on the community was not great. Indeed, it was the older residents who effectively defined the unspoken rules of the community, rules either about which the younger residents complained, or with which they complied. These rules were based upon how one should properly 'neighbour', operate a household, keep one's yard, define the boundaries between houses, sweep the sidewalk and care for one's garden.

Why should such domestic concerns have been so important to the older residents? They made the elders seem 'nosey' to the younger. I shall argue that the elders were highly motivated to share their values about work with the younger residents, as a way of ensuring the continued existence of their community. Values about work among the elders, expressed in old age, were formed during early, crucial stages in the personal development of these people that took place within a particular social formation[5] in which the work of coal-mining dominated both men's and women's daily work lives as well as their family lives. The economic conditions of company towns produced solidarity among neighbours, who depended upon each other to survive. The alliances formed among households were implicitly valued by the residents in later life and expressed to all members of the community through the fruits of their highly visible domestic work.

GROWING OLD

Old age is often conceptualized in biological or psychological terms, as though the ageing process were internally driven. But ageing is also a social process that can be analysed in terms of how people have actively constructed their lives, and how earlier life experiences and wider world events have shaped the conditions in which they live. The lives of older women in Appalachia in the 1970s immediately invite consideration of the historical circumstances of their youth in the late nineteenth or early twentieth century, and the recognition of the profound shifts in the position of women in the United States during that period of time. Although beyond the scope of this chapter, any discussion of Appalachia must make reference to coal-mining, the position that industry has held in the global economy and the fact that Appalachia itself has been 'underdeveloped' within the United States. These changing social conditions have filtered into Deckers Creek, affecting the very creation of the community and the changes it has undergone.

WHO WERE THE WEST VIRGINIA MINERS?

Most of the Deckers Creek residents were immigrants from Italy, Ireland, Hungary and Yugoslavia. Their presence in West Virginia in the 1920s and 1930s reflected an effort by that state begun in 1871 to recruit eastern Europeans to the mines (Bailey 1985). The recruitment of Europeans into the mines greatly inflated the population of West Virginia, which in 1880 stood at 3,701 and by 1917 had reached 88,665. Between 1890 and 1920 portions of the state had a fourfold increase in population; some counties had a ninefold increase.

The people who came to work as miners in West Virginia were both US mountain natives and immigrants, many of whom sought high wages. Table 8.1 illustrates the national origins of the miners during this time span.

While the table shows no Italian, Hungarian, Polish or Negro coal-miners in 1880, other sources indicate that a few were probably living

Table 8.1 National origins of West Virginia miners

	1880	1900	1911	1917
White*	2,777	12,028	30,094	48,237
Italian	–	554	8,184	7,388
Hungarian	–	–	4,106	4,346
Polish	–	220	2,181	2,117
English	447	1,053	505	–
Slav	–	–	1,841	1,215
Negro	–	4,620	11,950	18,128
Other	477	1,812	11,783†	7,234††
Total	3,701	20,287	70,644	88,665

* 'Whites' refers to native Appalachians.

† In 1911, this large number of 'other' miners was composed largely of Russian, Austrian, Romanian, Scottish and Irish.

†† In 1917, the 'other' miners were Austrian and Russian.

Source: Bailey (1985), p. 119.

in the state. Corbin (1981: 8) suggests that in 1880 there was a total of 924 European miners; by 1910 this had expanded to 28,000. The numbers of Austro-Hungarian and Slav miners increased after 1915, after Italy entered the First World War.

The older residents of Deckers Creek were part of this wave of immigration; many had arrived in West Virginia or nearby Pennsylvania as children. All the Deckers Creek elders from non-English speaking countries could speak their native languages, even though they had been living in the US since they were young children.

SPATIAL ARENAS OF SOCIAL RELATIONS

The social organization among the elders of Deckers Creek – marked by mutual support and close personal, neighbourly ties – developed in relation to the powerful presence of the coal-mine. When the mine was active (from the 1920s until 1950), the work women did was crucial to the men's ability to mine. Women kept wicks trimmed and clean for the lights on the men's hats, cleaned and mended clothes, provided kerosene and packed ample meals (the coal day for men began at 4.30 a.m. and ended at dusk). Before 1950, women did not work outside the home in waged jobs. After 1950, when the mine was shut down, some women took jobs in local shops, took in sewing or did cleaning.

Women's work was multifaceted and was wholly located within the boundaries of the community. Women worked above ground while the men worked below, and both had to work in order to survive.

The home: primary arena of work

The lack of a distinct border between the domestic and work worlds in the company town marks the extreme example that coal-mining communities provide in understanding the relationship between 'work' and 'home' in the United States. Unlike other occupations where, if a man loses his job, the family loses its home when it can no longer keep up the mortgage payments, in the coal-mining industry, a family automatically and unequivocally lost its home if a man lost his job. Since coal-mining was the dominant industry in Appalachia during the first half of the twentieth century, the residents of Deckers Creek, like other coal-mining families, had little choice about other jobs or living another sort of life.

The homes in which the residents of Deckers Creek lived in the 1920s were commonly referred to as coal shacks. Like most shacks built near the turn of the century, these were shabbily constructed, with raw beams, minimal or no plumbing and no bathrooms. They were heated by coal stoves which took a great deal of work to sustain. The houses

Plate 8.1 Kate stands on the sidewalk facing the coal route – along which trucks carrying coal pass every minute or so. Notice the row of houses and tree-lined street

provided minimal shelter. One retired miner from Deckers Creek described them this way:

'They repelled the rain most of the time. There was nothing to turn the cold except one layer of construction boards and many times these were not put together very tight, with the result that you waited until a warm day to take a bath. There was no running water in the houses. The water came from a pump that served at least four houses by means of a water pail. All the water used for cooking, bathing or washing clothes was carried into the house by the housewife, or if a family was lucky, by one of the boys. Fortunately, the water was usually good as far as purity was concerned, and was not the source of infections and diseases. Sometimes the kitchen had one electric light, which many times was supplied by the company and was hooked into the mine electric lines. So, when for any reason the mine did not work, there was no power to the houses. However, most of the lighting was done by kerosene lamps.'[6]

Equipped with few material resources, women spent their days fully engaged in demanding tasks, with their work structured around the

Plate 8.2 A young girl, the daughter of one of the coal-mining families, stands in front of her house in the late 1940s. She stands in the same spot where Kate (Plate 8.1) was photographed, thirty years later

rhythms of the mine. The following description of a woman's day was provided by Kate, who has lived in Deckers Creek since 1935. Her husband had been a boss at the mine; he trained some of the younger men in the community how to mine. She remembers what life was like above ground:

... when the men worked the mine during the 1930s. Early to bed, early to rise. In bed by 9.00, most of the time 8.00. Up at 4.00 a.m. Winter time (first thing) pot-bellied stove to get going. The fire had been banked (or covered) at night to keep. Open draft, shake ashes. Build the kitchen stove fire with kindling chopped during summer. Kids big enough did this chore. Cook breakfast. Slab bacon, eggs, fried potatoes was the menu. Pack lunch pail. Water in the bottom, sandwiches and fruit in the top. Apple mostly. Mix bread two mornings a week. There was no bakeries those days. Everything was baked at home, sometimes on the hearth outdoors. Get the kids up by 6.00. Feed, pack lunches, off to school. Make the beds, clean the stoves of ashes. The ashes were piled up in the back yard. After this, everything must be washed off, floor mopped. The evening meal was started mid-afternoon. When the husband came home determined the time to eat. Usually 5.00 to 7.00. In the summer the garden was cared for, picked and canned day-by-day. Nothing had to be store-bought but coffee, sugar and flour. Yeast was set with potatoes and water. After the evening meal dishes and cooking pots was washed and dried and put away. Dish towels (from flour sacks weighing 100 lbs), two large pans of hot water, bar of yellow homemade soap was the cleaning articles.

These spare words denote a frugal life. Nothing was wasted. It was the responsibility of the women to be able to manage on very little. They had great pride in their resourcefulness. Kate continued in her letter:

The clothes were made at home. Sewing was a big item. Comforts were made from worn woolens. Quilts were pieced from cotton scraps and quilted in large frames. Frames were up the year 'round. Sheets and pillow cases were made from brown muslin yardage. Every sheet had a seam down the middle. These were bleached with lye until white as snow. . . . Darning and mending was done in the evenings. Clothes were patched and repatched until nothing left to work on.

A woman's long work day was so heavily focused on the house that the term 'housewife', usually reserved for the domestic work of women, cannot adequately capture the nature of the labour she performed.[7] The workplace for women in coal-mining communities was her home, the location of survival and a source of pride in achieving it.

This pride was evident in the late 1970s. Some of the women still used the sheets and pillow cases that their mothers had made from bleached flour bags fifty years earlier. Indeed, the women frequently bragged to me that these very old items held up better than any sheets you could purchase today. At the wedding of one woman's daughter in 1978, several handmade, flour-bag, bleached pillow cases were collected from nearby households, washed and ironed, and slipped over the backs of the chairs at the wedding feast to keep everything perfectly white. The muslin pillow cases, once a necessity, became ritual symbols in celebrations, marking values developed during the mining days and expressed metaphorically through old material life.

The store: economic and political power

The residents had no control over the lack of plumbing or poor electricity, since the company had a monopoly on all sources of goods – from food, cigarettes and cloth to electricity, home and wages. The power that the company could exert through monopoly ownership of the mine and the houses in the community gave them obvious and enormous leverage. All the mining families I knew were acutely aware of this power and the degree to which they were exploited. Consciousness of their exploitation was accompanied by the pride they felt in their ability to survive and do good work.[8]

Men were paid at the company store in 'scrip', a form of company money. Wages were paid for each ton of coal dug (by hand, in low seams requiring the men to stoop as they dug). Prior to the formation of the union, the company had had the right to decide whether a ton of coal was 'contaminated' (containing slate in addition to coal). If so, the miner would be paid less than the full-ton rate. This, combined with the fact that the goods, fabrics and food came from the company store, and rent was paid there as well, meant that the mining family usually lived in perpetual debt to the company.[9]

Even in the face of company domination the residents were able to maintain tremendous solidarity:

'You know, so many things were scarce – coffee, sugar, yardage ... you couldn't beg a piece of muslin hardly. And he [the company store manager] used to get it in by the bagload up here, and the dirty devil would sell it out you know, and make a profit on it for himself, and the people here never got it. You just couldn't hardly buy anything. Well, he got kicked out because of it. I was there one day and I wanted some material, we used to wear little housedresses you know, and I wanted to sew up things like that for everyday things, and I was told through the underground that there was some that come in, you know. I went up there and he was there, and I asked about it, and he said, "no such a-thing, where did I get such an idea?" and all that business, you know. And, I don't remember, I said some pretty snotty things to him.'[10]

I asked her if she hadn't been afraid of the consequences of speaking back to the manager, given the power of the company in the town. 'Well no', she answered, 'cause I knew the butcher in the store.' The store butcher – who still lives in Deckers Creek today – was a neighbour.

Neighbourly relationships were powerful enough to give residents the strength to confront the company over such everyday matters. This mutual dependence among people within the community helped them to survive the exigencies of company town life. Depending upon one another served as a necessary buffer to the company owners. This dependence between neighbours, among men, among women and between spouses organized the relations between genders and within the household with greater solidarity than ordinarily found in other occupations.

A COMMUNITY IN TRANSITION: DECKERS CREEK AFTER THE CLOSING OF THE MINE

When the Deckers Creek mine shut down in 1951, everyone in the community was put out of work. When the company gave up the operation of the mine, it also gave up operating the community. It sold all the houses to an entrepreneur in a nearby city, who in turn sold them to the residents who lived in them. The relief of no longer living in a company town was compromised by having to scrounge for a living. The coal bust made it hard to find work. Many men found sporadic employment in other local mines, and others tried working as millhands or factory workers. Most, however, retired, either unable to find work or too debilitated by black lung and silicosis to continue. They lived on their pensions and social security. In 1969, when federal black lung benefits became available, this supplemented their income. Getting the federal black lung benefits was not easy: one not only had to have medical evidence of black lung, but had to prove continuous employment in the mines for a certain period of years. Since the mines had closed down and work was sporadic after 1950, this was no easy task. The fact that most men in the community did receive black lung benefits was due to the efforts of the women who 'dragged them down there' to sign up, and a local physician, who not only performed the medical verification for the presence of black lung or silicosis, but helped the men fill out forms which documented mine employment. Several of the men told me that they were not sure they were entitled to the federal benefits, since they already received some state benefits. It was the women – conscious of the effect upon themselves should they lose their men – who made sure the men received the federal benefits. Coal-mining wives always lived with the risk of losing their husbands. When they were young, death in the mine or the loss of job would result in their own loss of a home and community. In old age, death still loomed large, but the women were experienced in knowing how to collect the resources to help maintain themselves in widowhood.

Some women entered the work force for the first time during the

164

1950s. Two of them went to work in the local faucet factory, one worked in a dry-cleaning store and a few worked in homes, performing a variety of tasks such as hanging wallpaper and cleaning. One family moved to Baltimore, but found urban living such a strain that they returned to Deckers Creek after a year, preferring the more familiar if unpredictable life of mining, and retirement from mining.

Their attention focused on the community itself, the Deckers Creek families began to make their mark upon it. One way they did this was by improving their homes. The men plastered walls using hand-mixed plaster which cost little. They finished the raw beams. They built basements, put in bathrooms, installed plumbing, sump pumps and extra entrances. The fact that today these houses are so sturdy, neat and well-kept reflects the extensive domestic labour put into them since the 1950s when the mining families bought their homes. The work of improving the houses has never ceased. While I was there, one man built a new garage, another panelled and laid carpet in one section of his which he used as a workshop. Other families added aluminium siding, installed kitchen carpeting, repaned windows, rebuilt porches, fixed roofs, tarred driveways and refinished sidewalks in the two years of my study.

The closing of the mine in 1950 caused the men to refocus their work efforts. Making their houses comfortable for the long term represented the stake they claimed in the community which, for so long, had provided only tenuous shelter. With the closing of the mine and consequent retirement, the men's workplace moved from the mine to the home and community – where women had always worked. The intimate relation of home and work which had prevailed during mining days thus continued, but with a shift in the work culture.

DECKERS CREEK IN THE LATER 1970s

Domestic work completely defined the course of the day in Deckers Creek in the late 1970s. Too old to continue working in the mines, or too ill with arthritis, black lung or silicosis to perform any wage work whatsoever, the residents devoted their time and energies to keeping their town in shape and keeping in touch with each other. Tasks in and around the houses – gardening, keeping lawns mowed, hedges trimmed, producing tinned and frozen food – made up the day. There was a rough gender division of labour – men maintained the gardens, women did the tinning; men repaired furniture, women made covers for it; men trimmed hedges, women did laundry; men washed and painted the outside of their houses, women the inside. But some men helped to tin produce, and many women gardened. There was not, however, any conflict about the relative importance of men's or women's work; to the contrary, men and women tended to sympathize with how much the other had to do, often encouraging one another to slow down.

Although each household maintained a garden and took care of its own small piece of land, the residents shared the results of their efforts by exchanging them with one another.

Exchange

I was introduced to the system of exchange the day I moved into Deckers Creek. My first morning there I found a large shopping bag full of fresh vegetables on the back porch. More appeared over the next few days. Neighbours brought me salads, pasta dishes and pizzas. They lent me gardening tools, offered to exchange an old rug of mine for a washing machine of theirs and gave me seeds for my garden. Receiving such an abundance of gifts from my neighbours made me aware of the nature of exchange that occurred regularly among the older neighbours. During chats, visits and strolls down the alley, food or some other small token was frequently offered to one's neighbour, such as produce from the garden (a bunch of parsley, some fresh basil, one or two particularly choice tomatoes), a piece of cake or bread, or a handmade towel or pillow. Clearly not offered out of need – everyone had a garden – these gifts were shared to reinforce relationships that already existed. The exchanges of food-for-food, food-for-service and food-for-a-visit were routine.

Neighbouring

In addition to the reciprocity, routine comments illustrated the extent of interest that neighbours had in each other's gardens and houses. 'Oh, her flowers are always blooming and her lettuce grows all year, I don't know how she does it', or 'He just lets those tomatoes rot there, why doesn't he tie them up?', or 'She keeps such a clean house, I'm afraid to let her come into mine.' Neighbours took a generous interest in each other's work. The results of their work, their willingness to do it and to participate in exchanges resulting from it was, in this community, fair game for scrutiny of one another, and seemed to identify an individual's position within the community. 'Neighbourliness' in Deckers Creek was expected between people who lived next door to one another; less so between households separated by a few houses. 'Neighbouring' involved visiting, chatting, cooperating about how the back lawns should look, keeping one's garden weeded, exchanging produce and food and generally helping one another out. It involved communicating with one's neighbour, being available for aid, but not necessarily forming a friendship. The value placed on neighbouring well was most clearly seen in extreme situations. One neighbour, long an acquaintance of the older residents, had been mentally ill for a number of years. Her neighbouring took an exaggerated form. She knew the patterns well, and came to visit at odd moments (such as 8.00 in the morning), for odd reasons (to warn me, as she once did, that the devil lived next door); she swept the sidewalk – as did everyone – but did so for six hours at a time. She was treated kindly by her neighbours who had known her since childhood, although her 'madness' was little understood and occasionally feared.

A young woman who had moved into the community with her husband and one-year-old daughter was widely considered a poor neighbour. She did not speak when spoken to nor chat with her next-door neighbour, even when both were hanging laundry in the back yard at the same time. She was referred to (with some disdain) as 'Lady Smith' and her behaviour was likened to the woman who had gone mad. Even the young woman's dog, which ran up and down the street barking and snapping at passers-by, was considered to be 'mad'. Not to neighbour in Deckers Creek was indeed viewed by the older residents as demented behaviour.

Participating in neighbouring and exchange was a socially-organized community standard for Deckers Creek residents. Their exchanges reinforced both neighbourly ties and pride in work. These transactions wove the participants in a web of reciprocal relations that served as a tacit guarantee for aid from others during times of crisis; thus the gift exchanges were the warp and weft of a voluntary support system. The exchanges enabled the elders to depend upon one another in a graceful way, helping them avoid a dependence upon the social services provided by outsiders.

CONCLUSION

The lifetimes of these people have paralleled and reflected the great changes in technology, social organization, economy and patterns of work which have occurred over the past eighty years. Growing up and growing old during a century of dramatic social and industrial change has defined the pressing conditions of life for these ageing, former mining families. Forces external to them – such as the changing national demand for coal – altered the need for production in the coal-mine where they worked, and consequently changed the relationship of their community to the larger world. Over the course of the century their community shifted from being a company town, to a bedroom community, due to its altered relationship with the changing regional and national economy.

Until the late 1970s, when some older residents died, the population of Deckers Creek remained very stable. While children born in the 1940s and 1950s grew up and moved away, their parents remained, forging alternative work lives. The 'newcomers' who moved in became the focus of the elders' desire to share their meaning of the community. They also posed a symbolic and actual threat to the community as it had always been defined.

Work

It is common for work to be analysed in terms of production rather than morality or symbolic worth, in part because many analyses of work in the United States are made by economists rather than anthropologists. Multiple interpretations are credible, however, and make it clear that work itself is multidimensional: simultaneously economic, political and social. Conceptualizing work outside of a market-based definition expands the ways it can be analysed as a process and set of activities. Rather than focusing upon wages, surplus value, the internal labour market, the external labour market and the dual labour force as the only crucial features of work, one can begin to consider its social and cultural features. Looking at work without relying on a strict market approach is fruitful in understanding the work patterns which link people to a changing 'productive' society.

One of the best models for understanding work in its social context has been provided by feminists who have analysed women's domestic work in the home. This research has been extremely important in understanding the validity and value of women's domestic work by linking it to the workplace through the social reproduction of the labour force (see Benston 1969; Dalla Costa 1972; Secombe 1974; Edholm, Harris and Young 1978). The great value of this model, when applied to individuals who are not tied to the market at all, is in revealing the mutual connections between the realms of production and reproduction through the institutions of the family and marriage, and the domestic work of household members. The data presented in this chapter extend that feminist research by analysing women's exclusion from the workplace on the basis of age as well as gender. The data here show that domestic work is done by old women and men, not just by young and middle-aged 'housewives', revealing its connections and complementary ties with work in the production sphere.

Feminist research has enabled scholars to see that the work within the domestic and 'productive' domains are linked through the family. Much of that research, however, based the value of domestic work of women on their marital relationships to wage-earning men, and on the basis of motherhood (i.e. the biological and social reproduction of the labour force). Although domestic work was not considered part of the market economy, these analyses validated it through making clear its connections to the market economy, particularly via men's productive labour. Such analyses cannot account for the domestic situation of old women, divorced women or single women, whose work could not be considered to be 'valuable' since they lacked the vital connection of a spouse in the labour force. Further, these earlier feminist analyses focused on one 'slice' of the life course, and did not analyse how work might vary throughout life. The problem of understanding retirement in relation to production and social reproduction arises when confronting these questions. The multidimensional value of domestic work can be re-examined by looking at the social lives of people during retirement.

Using a developmental perspective towards work sheds light on social organization at several points in the life course and illuminates aspects of work other than the economic value of productive labour. Instead of validating domestic work solely in relation to the market economy, one can validate it in terms of the residents' perception of the meaning of informal, non-monetized or 'volitional' work. By adopting a non-market approach to work, and focusing upon the meaning of work for the residents themselves in relation to the shifting social circumstances in which they live, it has become apparent that the work of the older residents in Deckers Creek made them socially visible, both to them-selves and to the younger residents.

Studying a mining community has been especially fruitful for analy-sing the different domains of work. Mining is unusual as an occupation, because of the physical proximity between the workplace and home, the socio-economic and cultural ties between the two, and because miners tend to be a very conscious and politically active group of workers. The intimate relation between the coal-mine and the community, and the domination of the company over so many aspects of daily life there, affected the way in which the residents ordered their lives. They were able to resist control by the company through their relationship to home; their sense of belonging, usefulness and meaning emerged out of the solidarity they developed in opposition to the company. Although the conditions of the workplace affected the daily routine of work and the level of control that the worker had over the work process, it was within the community and within the household that many of the successes, frustrations, consciousness and attitudes about work were played out. Analysing labour processes clarifies the point that the awarenesses engendered in one domain may be enacted in the other, and that consciousness garnered during earlier life may come into play in a different form in later life.

Deckers Creek is a company town grown old; a town in which neighbours have lived and worked together for thirty or forty years and retired together as well. Work has involved not only mining, caring for one's garden, household and yard, it has included exchanging produce, cooked food and small handmade items after the mine closed. Participating in this system, defined and implemented by the older residents, has symbolized a way of taking care of one another and maintaining social relationships. Not to work was a sign of alienation from the community. Work functioned as 'grammatical knowledge' representing the shared experiences of having grown up and lived in a company town, and it served as a way of measuring one's moral stock in the community. Work served as a metaphor for one's position in the community, and understanding this was tacit knowledge for the older Deckers Creek residents.

The failure the elder residents of Deckers Creek experienced in trying to socialize their new, younger neighbours into the ways of the community was based on the fact that the newcomers had no clue about this tacit knowledge. This sort of knowledge is communicated through participating in activities over the years. Barring such shared experience, the elders sought to communicate it by inviting the newcomers into what seemed to them to be an obvious method of sharing: the exchange of produce. But the newcomers lacked the necessary reference that motivated exchange: being a coal-miner and living in a company town. The elders bolstered their efforts with fragments of stories about their past, mentioning what it took to roll the stick of dynamite, or go into the mine; stories the newcomers often mentioned having heard too often due to the poor memories of the elders. The hurdles to communication were the different referential frameworks within which the elders and newcomers operated. These frameworks represented changed socio-economic conditions that affected the work lives and workplaces of all these Appalachian residents.

The experience of older people becoming disenfranchised within their own community as a result of changed economic conditions and diminishing numbers reveals one social consequence of economic change. Shifts in residence induced by the new location of jobs produces communities without long lives, and social relationships without duration. The strength that the Deckers Creek elders found in one another was rooted in their old ties, a relationship their younger neighbours are unlikely ever to share.

9

LIFE COURSE AND SPACE

Dual careers and residential mobility among upper-middle-class families in the Île-de-France region

Jeanne Fagnani

INTRODUCTION

In France, as in other industrialized countries, the dramatically increasing participation of women in the labour force, leading to a growing number of two-income couples, has had profound repercussions upon every aspect of life, and particularly upon housing and residential location patterns. If the residential choices of these households and their impact upon urban forms and structures are to be better understood, explicit attention must be paid to the complex processes by which the locational decisions are made and to the range of factors that affect them (Stapleton 1980). Prior studies have shown that the presence of two earners in a household 'potentially both increases and narrows residential location options' (Hanson and Pratt 1988: 312) while the decision process becomes more complex, as additional parameters must be taken into account. Numerous studies on residential mobility have treated the household unit as an undifferentiated whole, however, and have not accounted for divergent needs and desires between male and female household members. Little empirical research has been conducted concerning housing and residential choice among husband and wife pairs (Michelson 1988); Shlay and Di Gregorio (1985), for example, demonstrated that women and men desire different neighbourhood worlds; however, they interviewed only one respondent per household.

As research is developed to examine divergence within the household, it is important to consider how the particular characteristics of couples and the contexts in which they make decisions shape their choices. In this chapter, I have chosen to examine a group that has emerged relatively recently in significant numbers in French life – upper-middle-class couples in which the adult woman is highly educated and employed. I focus on couples who mostly have two or three children, the wife being in early middle age. As Alice Rossi has noted, motives and aspirations can change over time and depend upon the stage of life (1985). The

171

couples I studied are all experiencing a transition point, having recently moved or being about to move residence within the Île-de-France region. My purpose is to analyse the negotiations, trade-offs, and compromises they make at this point in their lives as they seek to define and implement choices which will meet their personal and familial goals.

Background and hypotheses

The generation of women I am studying (aged between thirty-five and forty-two) has been strongly influenced by the feminist movement. As we have demonstrated in previous work (Castelain-Meunier and Fagnani 1988a, 1988b), these highly educated mothers, who understand the value of their degrees in the market-place, are eager to achieve all of their personal and professional goals. In their everyday lives, however, public and private spheres are fundamentally intertwined. On the one hand, in contrast with men, they do not forget their family obligations: they call home to speak with their children, they do some shopping at lunch-time, they try to adapt their work and family schedules. On the other hand, in order to meet their professional responsibilities, they will spend less time sleeping and devote less time to leisure than their partners, as time budget studies show. Nevertheless, the systematic relationship between the private and public spheres is not the only factor which bears on the choices they make. In order to analyse the strategies that these career-oriented mothers elaborate and their methods for constructing their lives within structural constraints, it is important to adopt an interactional approach to gender in these families. As Thompson and Walker (1989) state, gender in families includes structural constraints and opportunities, beliefs and ideology, actual arrangements and activities, meanings and experiences, diversity and change, and interaction and relation.

At this stage in their lives, these mothers are strongly committed to their jobs. Notwithstanding this, as a result of cultural dictates linked to the image of the 'good mother', of scheduling which does not recognize two-earner couples, and time/space constraints related to urban patterns in the Île-de-France, women face contradictory realities and must cope with conflicting obligations (Hantrais 1990). As a result of gender socialization, even if their partners divide the domestic and educational responsibilities equally, they carry most of the mental burden of responsibility for childcare: they plan, organize, delegate, supervise, and schedule. From this perspective, I have hypothesized that the choice of housing location plays a major role for these mothers in their professional planning and family management. As they try to muster all their forces to manage the dual obligations of home and work, the location of their residence, as well as the characteristics of their environment and psycho-social factors, affects the ways in which they manage their daily lives. The most salient factors influencing their residential choice should be proximity to the workplace; access to public transport (even if they have a car, they are often obliged to commute by public transport because they cannot park where they work, especially in central Paris (Fagnani 1986)); shopping; and day care for those having children under the age of three. These aspirations may be conflicting if they also want to offer their children a pleasant environment near to open green spaces. I have also hypothesized that these women are likely to play a major role in the residential choice process, because it represents a more important aspect of their lives than those of their partners, for whom career progress is the overwhelming priority. Moreover, these women want to realize their ideal of a loving marriage. In this context, they often prefer to give up the idea of struggling with their partner over the sharing of domestic chores. In exchange, and implicitly, their husbands give them priority in housing choices. Therefore it seems likely that the couple's residential choice will be influenced by the location of the woman's workplace.

A new balance of power is created in these couples by the endowment of the woman partner with professional and cultural assets, and an ideological context in which equality between the sexes is increasingly becoming the implicit regulation and the cultural norm. Educationally advantaged women have assets and resources with which to influence significantly the process of decision-making about residential and housing choice, among other things. At the same time, in the normal give-and-take of marital interaction, and in the interests of preserving 'marital peace', women are obliged to take into account the needs and wishes of their spouse in order to achieve a consensus. In addition, they want their husbands to spend some time with the children and to be involved in fathering. Thus, in studying residential and housing choices, I will pay particular attention to the ways in which decisions are influenced by alternative ideas about family arrangements. Further, I will analyse the housing decisions in terms of location choices: city centre (Paris *intra muros*) versus suburban location (near or outlying suburbs). These are quite distinct environments in Paris. The division between the city and the suburbs is particularly marked by an elevated *auto-route* which encircles the city and defines its limits. The two spatial contexts offer contrasting lifestyle possibilities. Thus choices are likely to reflect attitudes both towards family and towards society.

Methodology

I adopted two complementary approaches designed to identify decision-making processes within the upper-middle-class households and to set these households in the larger Parisian context. I conducted in-depth interviews with forty mothers and twenty of their partners, and analysed data collected in the housing inquiry of 1988 by INSEE (Institut National de la Statistique et des Études Économiques). All the women and partners interviewed were between thirty-five and forty-five years of age and held high-level management or administrative positions or professional posts (including teachers). All had attended at least four years of high school and some had completed a higher degree.

The couples were selected by using the 'snowball' method: I knew seventeen of the couples personally and this enabled me to gain their confidence and thus obtain valuable information which would otherwise have been difficult to obtain. All couples had moved within a four-year period or were, at the time of the survey, looking for new housing. Half of the families were living in the centre; the other half in the near or outlying suburbs. The husband and wife were interviewed separately in the home. Each couple in search of new housing participated in a series of interviews, each of which coincided with a step in their search process. Including repeat interviews in some households, altogether I completed seventy-two interviews. The insight gained into subjective dilemmas and processes involved in locational decisions was of an order not often resulting from large statistical studies.

I also examined data from the housing inquiry of 1988 conducted by INSEE, which included a random sample of 35,500 housing units, of which 11,000 were in the Île-de-France region. From this larger data set I extracted material on the upper-middle-class couples living in the region (those whose head of household was a professional or in a high-level managerial position) to undertake a secondary analysis of the

Table 9.1 Some housing and demographic characteristics in each area of the Île-de-France region

Place of residence	Average number of rooms per dwelling	% of single-family homes	% of renters	% of home owners	% of households with 4 persons or more	% of single-person households
Paris	2.6	0.9	59.3	28.0	12.4	44.5
Near suburbs (Area 2)	3.2	21.3	56.0	39.0	25.1	27.0
Outlying suburbs (Area 3)	3.8	39.5	45.5	50.0	30.9	21.2
Peripheral areas (Area 4)	4.1	66.9	30.2	65.0	34.3	16.0

Source: Housing inquiry 1988, INSEE.

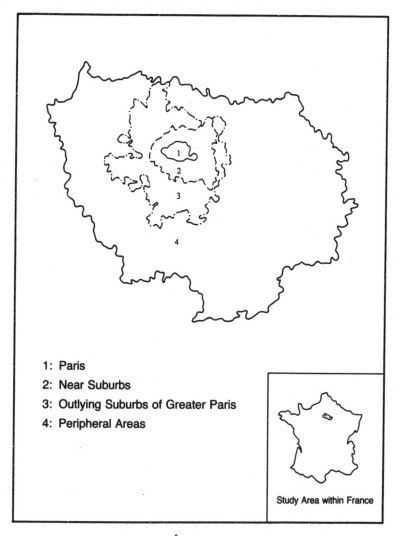

1: Paris
2: Near Suburbs
3: Outlying Suburbs of Greater Paris
4: Peripheral Areas

Study Area within France

Figure 9.1 Île-de-France region

respective commuting times of men and women within the study population, according to their place of residence.

The social sub-group examined in my study is especially interesting because, unlike less privileged sub-groups, they benefit from a certain freedom of choice. Thanks to their dual incomes, they have more freedom to manoeuvre in the property market.

HOUSING PATTERNS AND DEMOGRAPHIC SEGREGATION IN PARIS: A RADIO-CONCENTRIC PATTERN

As Hanson and Pratt (1988: 305) have stated, 'individuals' decisions not only shape urban form . . . such decisions are also shaped by the urban context in which people find themselves'. It is necessary, therefore, to describe briefly the specific spatial context of the Île-de-France region before analysing individual decisions.

The Île-de-France can be divided into four statistically concentric areas (Figure 9.1) which radiate from the centre of Paris and comprise the whole region. Taking these areas as a starting point, we can observe the radio-concentric patterns of demographic segregation and housing characteristics (Table 9.1). Some studies also demonstrate that distance from the centre of the city has an important impact on household and housing patterns (Chauviré 1988; Team 'Espace, Population, Société' 1989). This phenomenon suggests a relationship between life course stages and living space. As a matter of fact, families with young children are preponderant among the households moving into the outlying suburbs. Among those moving into central Paris, young adults and single-person (economically active) households are overrepresented (Bonvalet and Fribourg 1990).

In central Paris, the multi-family rental sector dominates total housing stock, 58 per cent of which is composed of one- or two-room apartment units, while in outlying suburbs and peripheral areas, single-family dwellings, most often owner-occupied, predominate. The near suburbs have a more balanced housing stock in terms of type and tenure of housing.

As one might expect, single-person households are overrepresented in the city centre (Chauviré 1988). Most of these single-person Parisian households are female (61 per cent). The number of households with three persons or more, expressed as a percentage of total households, increases within each area from the centre to the periphery. For example, 16.6 per cent of married women aged under forty with one child and 11.4 per cent of those with two children live in the city centre, compared to 52 per cent of women aged under forty living alone and without children.

Two important factors provide a partial explanation for these phenomena: the diminution of housing costs from the centre to the periphery[1] and the growing desire of families to become home-owners (Taffin 1987). The highest housing costs, either for renting or buying, are found in central Paris, which is affected by ongoing gentrification (Bessy 1987).[2] Data from a 1988 survey of a representative sample of purchasers of second-hand dwellings in Paris (Massot 1990) show that the group comprised of entrepreneurs, executives, and members of liberal professions acquired half of such dwellings placed on the market. Their purchases amounted to two-thirds of the total annual turnover on the housing market, reflecting the fact that they 'trade up', buying more expensive apartments that are often bigger, more comfortable, and of a better than average quality. Furthermore, these apartments are often located in the most highly-rated neighbourhoods.

Such prices, of course, put families with children and low- to middle-income households at a disadvantage. As a result, between 1975 and 1982, the number of dual-earner couples within central Paris decreased by 8.9 per cent, while in outlying suburbs and peripheral areas they increased by 24.9 per cent (Le Jeannic 1990). Moreover, there was a spectacular rise in the sale prices of houses in the suburban areas best served by underground and commuter trains, which also penalized low-income families.

Whatever their social status, more and more Parisian families want to become home-owners, for both ideological and financial reasons. The desire to own property goes hand in hand with the growing desire to have a home of one's own. Indeed, the dream of a single-family home is very strong. It is part of a self-perception that is tied to a 'private world', protected from the 'dangers' of city life and adapted to children's needs and the family-oriented life. A house represents a legacy for one's children and takes away the pressure of spiralling rents in the region. In 1988, 42 per cent of households were comprised of home-owners, compared with only 36.4 per cent in 1975. However most families are obliged to move to the suburbs to become home-owners; Table 9.1 shows clearly that the percentage of home-ownership increases from the centre to the periphery.

Within this spatial context, even the couples in the study (whose monthly incomes varied, before taxes, in the range US$5–10,000) must choose among and rank their priorities. They find themselves caught up in a process of trade-offs in which some aspirations are inevitably given up, either in the quality of housing or in the quality of location. If they want to live in Paris, they must either give up the idea of becoming owners or they must buy a smaller dwelling than that which they might afford in the suburbs. Alternatively, if they want to own a large comfortable dwelling or a single-family house, they must accept a location much further away from the centre. As one woman stated, 'One can't have everything!'

RESIDENTIAL CHOICES, HOME–WORK LINKS, AND CONJUGAL INTERACTIONS

Reasons for moving

Data from the housing inquiry survey show that the primary reason for moving is the desire for a larger dwelling. The second most frequent reason is the desire for home-ownership. Nevertheless, there are often multiple reasons for mobility. The desire for a change of lifestyle can result in a move from Paris to the suburbs, or the reverse, while simultaneously reducing travel time to work and/or facilitating the acquisition of a larger living space. For couples within the study, these desires are often realized following pay rises or as a result of inheritances or financial assistance from the family. Financial assistance from relatives in order to buy a home is a current practice in all French social groups (Gotman 1989). The partner whose family contributes to the purchase gains major bargaining power in the couple's housing search.

One criterion often predominates in housing choice: a room for each child. This standard corresponds to the values and educational practices of the middle-class social milieu, in particular, the respect for individual autonomy, the need to be alone, and the desire to have one's own space (Neitzert 1990). As one woman said, 'children must learn to manage their own space as they will later learn to manage their own lives'. However, taking into account the high cost of housing within Paris, this requirement often produces a dilemma for couples living in the centre and expecting a second or third child. Either they can move to a larger house in the suburbs, one better suited to the needs of a growing family but less convenient to their daily lives, or they can give up a certain degree of comfort and continue to live in the centre near their jobs, in an environment better suited to their current lifestyle. As we will see below, a myriad of trade-offs in family organization must be made when choosing a location and a dwelling: size, living style, whether individual or collective, owner or rental status, and the respective lifestyles of each partner, as well as their respective commuting times.

Commuting time is often shorter for women partners

Commuting time is often shorter for women than for their partners, a phenomenon which mirrors the gender-based family organization. All study respondents, men and women, had selected their current job before they had chosen their dwelling. As Hanson and Pratt (1988: 311) have stated, 'It is particularly for professionals and managers . . . that employment location is likely to influence residential choice' and 'professional women are less likely than are other groups of employed women to select their residential location before their job' (Hanson and Pratt 1988). However, among our couples, it is not uncommon that, once moved, women request a site transfer within the same company or organization in order to reduce commuting time. In fact, many of these French professional women interviewed work either in schools or public services attached to local administrations, as teachers, doctors, social workers, or in government administration, which enables them to find work more easily in the local job market.

For these couples, proximity to the worksite is one of the dimensions of the quality of life to which they aspire. Long commuting times are perceived as an undesirable waste of time. However, reducing commuting time for both partners is problematic, since the spatial distribution of employment is clearly gender-segregated. Overall, women's employment is more city-centre oriented than is men's employment.[3]

Because these mothers are concerned about spending some time with their children in the evenings after work (or feel guilty if they do not) and because they also often carry most of the burden of childcare, for them, reducing commuting time is crucial. Their partners, caught up in their work, are often reluctant to carry an equal share of family responsibilities, or cannot do so. These men often feel obliged, in exchange, to accept a locational trade-off which favours their spouse.

Results from several studies show that women work closer to home than men: in the Île-de-France region, among dual-earner couples, the average distance to work for women is 7.4 km compared to 10.9 km for men (Le Jeannic 1990). There are twice as many couples in which the woman works near the home (i.e. in her residential community or in a neighbouring one) and the man further away, than the opposite. Data from the housing inquiry study also show that, among dual-earner couples whose head of household is a professional or an executive, women have shorter commuting times than their partners, irrespective of the place of residence – in the whole region, men travelled on average 36.3 minutes to work whereas women averaged 29.9 minutes. For Paris residents, these travel times are 31.5 and 27.8 minutes respectively. Within our sample of forty couples, twenty-five women have shorter commuting times than their partners. These data accord with those reported in studies of other urban contexts (Blumen and Kellerman 1990; Hanson and Johnson 1985; Hanson and Pratt 1990; Johnson-Anumonwo 1988; Madden 1981; O'Donnel and Stueve 1981; Villeneuve and Rose 1988).

Obviously, for the couples having the same or adjacent places of work, a consensus is easily achieved. The recent study by Le Jeannic (1990) shows that this is more frequently the case for couples living in Paris than for those in the suburbs. Of all dual-income couples, both partners in 29.7 per cent of those couples residing in Paris work close to home, compared with 24 per cent for those living in the whole Île-de-France region.

Within the couples of our sample, however, who are beginning to adopt the values of 'new parenthood' (i.e where the father and mother share (or try to share) equally the burden of child-rearing), 25 per cent have chosen a residential location allowing the same commuting time for each of the parents (ten of forty couples). This compromise reflects the 'search for equity' and the real desire of some men to be involved with fathering.

Finally, the women who have, despite everything, accepted a longer commute (five of forty), are those whose two priorities were first, residence in an outlying suburb for the good of their children (see below), and second, to become home-owners. These women are likely to engage a cleaning woman or family aide to do domestic chores and/ or are able to rely on a relative (usually their own mother) to care for their children at home or after school.

VALUE-SYSTEMS AND LIFESTYLES: DILEMMAS, TRADE-OFFS, AND COMPROMISES

Over and above the diverse criteria and frequent compromises that contribute to the decision about housing and residential location, one fact stands out: men frequently accept the trade-offs suggested by their wives and give in to what they rationalize as their needs. This reflects an implicit understanding that men recognize that women still orchestrate family life. These mothers have been able to advocate for their specific household considerations when negotiating with their partners. Not surprisingly, therefore, the woman, more often than not, is charged with finding a new place to live. Disagreements among couples may occur: the wife and the husband may hold differing views on the advantages or disadvantages of particular locations relative to other options, particularly when the mother has different ideas about the education of their children than the father, or when the lifestyle aspirations of the two partners differ. The compromise they arrive at depends upon the abilities and assets of each, as well as the division of power within the couple.

Choosing the suburbs and a family-oriented lifestyle

Moving to or within the suburbs proved to be a choice for an environment adapted to a family-oriented lifestyle. It is also sometimes part of preparing for a larger family. For the most part, suburban immigrants become owners of a house with a garden; in moving out of the city centre they wanted to offer their children a 'safe' and pleasant environment, near to open green spaces.[4]

Mothers like the lifestyle in the outlying suburbs more when they work part-time (until children reach school age) rather than full-time, or if they have jobs less than thirty minutes' travel from home. This situation allows them to safeguard the flexibility of their time-management. Sometimes, they have sacrificed their preference for urban living and accepted long journeys to work in order to move to a more suburban environment which seems more conducive to child-rearing. There is, of course, some underlying truth in their assessment: inner city areas of Paris are often unsafe or poorly suited for children's play, while suburban areas offer greater outdoor play opportunities. Some suburban women, however, work in Paris, and are faced with long commutes either on public transport (because they have no parking place where they work) or by car. These women say they have accepted this arrangement in exchange for a 'more agreeable environment for the children and the family'. They have a more comfortable house, where 'everyone has their own space'. They do not mention the attractions of the city, but rather its disadvantages: the noise, the crowds, the lack of space, and so on – as if they sometimes want to convince themselves they made the right choice. The regret of having left Paris, however, sometimes comes through in conversations with women and men who moved from the centre to the suburbs. They stress the fact that they are far away from cultural amenities, that they sometimes feel isolated (as one woman said, 'our friends visit us less frequently than before'), or that they have to commute every day to Paris, which is time-consuming and tiring. All these couples, however, have a very family-oriented lifestyle. For example, they go out in the evening less frequently than couples who live in, or near, the city centre.

Even if mothering does not represent the ultimate fulfilment to which they could devote their lives, the women who preferred (and sometimes struggled) to move to the outlying suburbs, have strongly internalized images of the 'good mother'. Furthermore, since they benefit from the assistance of a paid family helper, the struggle over the sharing of domestic chores becomes secondary. They do not use their power to push for a fair distribution of labour at home;[5] on the contrary, they are more preoccupied with preserving 'conjugal peace' and with allowing the career trajectory of their husbands to remain unhampered. As in the productive sphere, so too they are able to develop creative responses in the reproductive sphere. As other researchers have stated (Pleck 1985), these high-achiever mothers often only want their husbands to be more involved with fathering.

The move to or within the suburbs can also coincide with the mother's desire to devote herself temporarily to her young children and to take a break in her own occupational mobility by taking a leave or working part-time. But even if their children are of school age, they have educational needs that are demanding of the mother's time. The women feel the need to spend time with their teenaged children and put value on the 'quality' of time spent with them. Sometimes, they even come back home during the weekday to have lunch with them. Still, these women find aspects of suburban living burdensome. In residential areas where single-family homes predominate, housekeeping is often made difficult by the long distances from home to services, schools, and shops. The women often complained about the need to chauffeur their children everywhere in the car.

Research in North America suggests that although women find the suburbs less satisfying for their individual opportunities, they find them more amenable for contented family living (Michelson 1977; Saegert 1981). This distinction was not always the case among the suburban dwelling women in our sample. The proximity of their workplaces and of small suburban centres well equipped with services and shops (such as exist in the inner suburbs, for example in the 'new towns')[6] and the presence in some cases of relatives, allowed them to maximize both opportunities for themselves and the care for their children. Furthermore, by acquiring a larger dwelling, they have been able to satisfy their desire to have 'a space of my own', which Parisian women often must renounce. Conversely their husbands complain when they have long commutes, especially if they espouse the image of the 'new fatherhood' which is emerging in this social sub-group and wish to spend more time with their children in the evening.

The experiences of two couples reveal the marked influence exercised by the family organizational pattern on the decision to migrate from Paris to the suburbs, in order to satisfy the desire for home-ownership. The first couple had been renting a three-room apartment in a central area of Paris. They bought a single-family home in the inner suburbs in order 'to grow and become home-owners', because they could not afford to purchase a larger dwelling in the areas of Paris they preferred. This relocation seemed to them a reasonable compromise, because 'each of us has the same commuting time [fifteen minutes by car], and the children are happy here with a garden'. Even though the woman of this couple (a researcher in a public centre, with a flexible work schedule) orchestrates and supervises the daily organization of the family, the father (a high-level manager in a private enterprise) is strongly involved in the educational duties, and they cooperate to assume responsibility for care of their two children if one of them is obliged to travel or to return home late in the evening.

The trade-offs were much less easy to achieve for the second couple, whose places of work were situated on opposite sides of the Greater Paris area (the man is an executive in a company in the western inner suburbs, his wife, a secondary school teacher in the eastern suburbs). Each favoured a different lifestyle. The man demonstrated a strong taste for city living – he liked going out in the evening and appreciated the proximity of cultural events and the animated ambiance of his neighbourhood. Since the birth of their two children (aged six and eight years), the woman has wished for a calmer, more family-oriented life and is less willing to tolerate the constraints of commuting (forty-five minutes' travel to work).

The lack of space for growing children, the woman's wish for a room of her own, anticipated increases in rents, and the possibility of home-ownership, led the couple to consider moving. The man, reluctant to move to the suburbs, tried to convince his wife to stay in Paris, but 'there was always something that she didn't like in the apartments we visited, or it was the neighbourhood, too noisy or too ugly'. Additionally, the fact that the woman had always taken more care of the children than had her husband legitimated her right (in her own eyes as well as those of her guilt-ridden husband) to demand housing located close to her place of work. Strengthened by financial assistance from her family, she carried the decision and chose an apartment in the new town of Marne-la-Vallée, in the eastern suburbs of Paris. Mixed land-use patterns in this area and proximity to work allow her to combine her domestic and professional commitments more easily. Her partner, on the other hand, complains of his lengthened commuting time and expresses undiminishing regret at having left Paris.

Living in town at any price: the attachment to urban values and the subordination of housing to the choice of location

The most prevalent attitude among couples who live in Paris is a rejection of suburban life and a 'sentimental' attachment to the city centre. Both men and women mention the advantages of living centrally, but it is the women who underline the practicalities: proximity to shops and services, public transport, daycare, kindergartens, schools, and so on, and for most of them, proximity to jobs. Women who have young children, however, often made remarks such as 'Paris isn't a place for young children . . . the city isn't made for them.' It is as if these mothers felt a gap or a contradiction between their own needs and those of their young children. This contradiction was less frequently felt by mothers who had teenaged children. These women registered the advantages of independence, the cultural wealth of the capital, and access to the most prestigious schools.

Couples who are particularly social or who have a very 'open' way of life appear to be the ones who chose to live in Paris, despite the arrival of children in their lives. These are couples who go out a lot in the evening (clearly much more than those who live in the far suburbs) and who take advantage of the cultural opportunities of the capital such as theatres, cinemas, museums, and art exhibitions. Those who have always lived in Paris are also concerned about preserving neighbourhood ties to friends and relatives. Although several of the women complained about the cramped nature of their housing, they nevertheless appeared more concerned about maintaining their current lifestyle, and seemed reluctant to change their daily routines.

Interviews with couples with two children who live in Paris show what an influence the problem of living conditions and the constraints of the housing market have on their decision to have another child. A third child would force or had forced certain couples to look for a larger dwelling. In most cases, this meant a move to the suburbs or even to peripheral areas. Mothers and their partners contemplated the prospect of a completely new lifestyle with much trepidation. In the end, a certain attachment for their neighbourhood, or a rejection of the 'suburban life', contributed to their decision not to have a third child (Fagnani 1990).

Thus, households in Paris were explicit about their decision to accept lower housing quality and size, and, in some cases, to forgo home-ownership, in order either to remain where they had always lived or to come back to the centre where they had lived before having children. In Paris, most of these families are renters: this is not necessarily a reflection of preferences for this type of tenure, but rather that they cannot afford to purchase, especially if they want to devote part of their income to leisure, entertainment, travel, and a full-time paid family helper.

Parisian parents insisted more often than those living in the suburbs on the 'need for autonomy' of their adolescent children: 'they can get about alone, go everywhere on foot because everything is nearby'. They have access to the best schools, which responds to the concerns of parents about their children's social success. When they have young children (under six years of age), the need for access to all-day care or after-school care facilities is another factor which contributes to the desire to live in town. Daycare centres are more numerous in Paris and in the near suburbs than in the outer areas (Pinçon, Pinçon-Charlot and Rendu 1986).

The following case illustrates the kind of negotiations that occur within couples where women demonstrate a strong taste for city living when their children get older.

Until 1987, the woman (forty-two years old), an executive in a public administration, and her partner, a researcher in a public institution, rented a single-family house in an attractive residential area located in the southern outlying suburbs:

'The children were very happy there. Paul also, he was only ten minutes from work by car . . . but me, it took me at least forty-five minutes to get to Paris [where she works] and I was fed up with living in the suburbs.'

After 1986, their two children having grown (the older was eighteen, the second was seven), she wanted more and more strongly to go back to Paris to live.

> 'I wanted a different lifestyle, to be able to go out more easily in the evening, to have everything close by without taking the car . . . and I didn't want to be obliged to buy a motor scooter for my son so that he could get about, that scared me too much . . . It took me a year to convince them!'

The most reluctant were the children, who 'didn't want to leave their friends'. She had much less trouble convincing her husband who 'didn't object' and who, very interested in his work, seemed to leave the task of orchestrating family life to her. In addition, she had shown him the advantages of being able to enrol their children in prestigious schools.

> 'He was very sweet about it, but I am the one who took care of everything . . . I looked for an apartment near a subway station so that he could go to work by metro without changing trains.'

In 1987, the family moved into a three-bedroom apartment located in the 5th *arrondissement*, with a monthly rent of approximately US$1365 (8,200 FF, fees included).

> 'Buying an apartment was out of the question. We could not afford it but we didn't care. I inherited an apartment in Vesinet from my parents [in the western suburbs of Paris] which I rent. It makes up for the rent here.'

The comforts they have given up are largely compensated for in her eyes (and now in those of the other family members) by the advantages of the location. Even the high rent is minimized by them: 'we have reduced our transport expenses significantly and we have sold the second car!' Even though this woman had been the catalyst in the decision to return to the centre, that decision was associated with the common desire of the couple to recommence their marital dynamic and to establish a new kind of relationship based on the autonomy of each.

CONCLUSION

These couples, basically in the same stage of the female life course and with comparable socio-demographic characteristics, reveal commonalities. With two professional incomes, their status makes it financially possible for them to have some choice with regard to housing. However, this same condition sets up tensions between career and family priorities for both partners. They are at the life stage when their children are living at home and are dependent on them in a variety of ways – for example, they need space, transportation, supervision. This circumstance sets up certain housing needs, especially when each member of the household wants their own space. All these women are concerned about the achievement of their different goals and show a great deal of ability in organizing their family life.

However, whatever the age of their children, they find themselves at a period of their life course when the structural conflicts between being strongly committed to their job and educating children become most acute. This is why they are so sensitive about the residential and housing issues. The age of their children has an impact on the couples' residential choices. When their children are very young or are under school age, women tend (or are likely) to give them priorities over their own aspirations: they prefer to live in suburban areas better suited to their children's needs, even if it sometimes complicates their daily organization and obliges their husbands to commute long distances. When their children are teenagers, those couples who demonstrate a strong taste for city living can more easily fulfil their aspirations and move to the centre (where they rent or buy an apartment if they can afford to) because they put value on the advantages of living in central Paris for their children: autonomy, cultural wealth, and access to the best schools. But these couples also differ in their residential preferences and expectations according to their value systems and the respective strategies both partners try to elaborate in the private and public spheres. In examining their residential choices, we see how a group of educationally advantaged mothers displays a varied range of responses to the structural dilemmas facing all women as they make efforts to cope with life in an urban society where social values and institutions still place the greatest responsibilities for family care on women, even though values about fathers' roles are changing to a degree, and where spatial and temporal constraints make it difficult for women to combine family and work roles. The study also reflects the increasing variety of alternatives to family living arrangements which are developing within subpopulations, and questions the traditional theories of residential mobility as well as traditional distinctions between public and private spheres.

What might be the long-term implications of the housing choices made at this point in the life course for their later years? It is interesting to note that those who chose to buy a large house in the suburbs contemplate returning to the centre when their children have left home. With this objective in mind, those who can afford it buy a small apartment in the centre, which they either rent out or use as a place for one of their children to live. The others think they will sell their house in the future when their housing needs have changed and then move closer to the centre. These attitudes reflect current patterns of residential mobility in the Île-de-France (Bonvalet and Fribourg 1990) and it is likely that they are contributing to the continuing gentrification of central Paris. Indeed, between 1978 and 1988 households of high-level executives and professionals increased from 17 to 27 per cent of the total living in Paris. One might expect that the increasing diversity of family structures and the growing desire of households to adjust their housing needs at every stage of their life course will lead to increasing residential mobility.

Thus, if urban residential theories are to explain fully the functioning of current urban societies, it is not only necessary to develop a more flexible view of families, both as sources of housing demand and as actors in the housing market (Stapleton 1980), it is also necessary to reassess the nature of home–work links (Hanson and Pratt 1988) *for each partner* within dual-earner couples and to consider their residential choice processes as an outcome of new gendered power relationships whose patterns have changed considerably in the last three decades.

10

LOCAL CHILDCARE STRATEGIES IN MONTRÉAL, QUÉBEC

The mediations of state policies, class and ethnicity in the life courses of families with young children

Damaris Rose

During the past decade or so in Canada, the number of places in organized child daycare services has increased considerably, both in absolute numbers and relative to the number of mothers in the labour force. Nevertheless, access to these services remains very restricted and, for most parents, expensive. Most children are cared for 'informally' – by unlicensed caregivers; by siblings; by friends; by relatives; by the other parent (the two parents frequently working different shifts in order to arrange childcare); or by themselves.

Privatization and deregulation of service provision have begun to gain ground in Canada (especially since the implementation of the free trade agreement with the USA), reinforcing longer-standing monetarist pressures on federal and provincial governments to reduce deficits by cutting social spending (Lightman and Irving 1991). Consequently, for example, the share of daycare centres' operating expenses paid for by the State, already meagre compared to European countries such as France or Denmark (see, for example, Hatchuel 1989; Phillips and Moss 1989) is being reduced further. Concomitantly, a greater proportion of the limited funding available for childcare is now channelled to parents through income tax deductions that can offset the costs of informal as well as organized modes of care; supposedly, this is to increase parents' freedom of choice, although the reality beyond the rhetoric is very different.

Meanwhile, 'family policies' (see, for example, Québec, Comité ministériel permanent du développement social 1984) aim to increase the flexibility of the arrangements families can make to combine paid work and childcare by emphasizing care within the family, yet without concomitantly fostering changes in the workplace, such as elimination of overtime requirements and extension of subsidized workplace child-care,[1] or making more than token moves to extend the existing very limited statutory paid parental leave provisions (see Canada, National Council on the Status of Women 1986).[2]

This chapter presents a conceptually- and historically-grounded account of state policy concerning childcare in the province of Québec, and examines how access to the current array of 'organized' services and 'informal' childcare arrangements is determined in three Montréal neighbourhoods. Focusing on Canadian-born and immigrant families with young children needing pre-school and before- and after-school childcare, it explores how the ways that these parents arrange childcare reflect the interplay of life course position with class, ethnicity and family structure. The options they have and the choices they make are situated within the parameters of the wider politico-institutional framework as this is mediated by socio-economic and cultural factors which vary from one neighbourhood to another.

A FRAMEWORK FOR EXAMINING MODES OF CHILDCARE PROVISION

To make sense of the historical development of childcare policy in Québec and how different modes of provision can affect access to services by different groups, we need ways of conceptualizing childcare provision.[3] Rather than developing a taxonomic classification (cf. Mark-Lawson, Savage and Warde 1985: 196), it seems more useful to propose four major dimensions of analysis, of which more than one may be needed to describe a particular mode of care (Table 10.1).

First, the distinction between organized or formal and non-organized or informal forms of care (see Harrison 1986) not only raises the question of whether the State is involved, but also whether parents can seek out an already-existing service offered to a group (e.g. a daycare centre) or whether the service only exists once it is negotiated between parent and caregiver (e.g. a babysitter who comes to the home). This inevitably has implications for whether the service is 'rigid' or 'flexible' in terms of the hours of service, its content and the age groups involved.

The second dimension distinguishes 'market' and 'non-market' pro-vision of group daycare services, according to whether an operating subsidy is received (it is important to bear in mind, however, that users of market forms may still be subsidized by grants or income tax allowances). A service can also be non-market in a more fundamental way, for example, care by a grandparent whose role includes passing on family values, traditional cultural practices or the language of origin. Moreover – and this brings us to the third dimension – the market/

Table 10.1 Conceptualizing modes of childcare provision: dimensions of analysis applied to the Québec case

Organized, 'formal', rigid services	Unorganized, 'informal', flexible services

– offered to a group
– hours fixed
– content regulated
– state licensed
  eg: daycare centres
    school daycare
    licensed family home daycare
	– existence of service is negotiated on individual 'contract' basis between parents and care-providers
	– State not involved
	  eg: unlicensed caregivers
	    care by relatives

Market provision	**Non-market provision**

– no operating subsidy
  eg: commercial daycare centres
    unlicensed family home daycare
	– receive operating subsidy
	  eg: coop non-profit care (parent-controlled)
	    licensed home daycare
	    school daycare
	– operate without market criteria
	  eg: care by sibling, friend, relative
	    parent care

For profit	**Non-profit**

eg: commercial daycare
  licensed home daycare
  babysitters paid market rates
  some workplace daycare
	eg: municipal daycare
	  school daycare
	  coop daycare (parent-controlled)
	  care by relatives or friends, provided free or at below-market rates
	  some workplace daycare

Services requiring paid labour only	**Services partly or fully dependent on volunteer labour**

eg: commercial daycare centres
  licensed and unlicensed family home daycare and babysitters
	– organized services requiring parent involvement to set them up:
	  eg: coop non-profit care
	    school daycare
	– organized services requiring parent involvement to help run them:
	  eg: coop non-profit daycare (management, maintenance)
	    school daycare (sometimes has partly volunteer staff)
	– care by relatives, provided free

non-market distinction needs to be considered in tandem with that of for-profit/non-profit. Some services may be subsidized and yet be for-profit (e.g. small licensed home daycare businesses); other subsidized services are non-profit (e.g. parent-run coop daycare). In Canada, commercial centres generally offer a lower quality of care (higher staff turnover, more breaches of safety regulations, etc.) than non-profit centres (see Canada, National Council on the Status of Women 1986: 49–51; Prentice 1989).[4]

The fourth dimension shown in Table 10.1 – services requiring paid labour versus those partly or fully dependent on volunteer labour – is a key one, indicating that the so-called 'informal sector' includes both paid and unpaid caregiving work. Notably, 'granny's' services are not necessarily always free, nor is her availability to be taken for granted: for instance, depending on her position in the life course and her economic situation, she may still be in paid employment. With respect to organized services, this dimension highlights the fact that, in some jurisdictions (notably Québec), the very existence of the formal system may depend largely on local parents' capacities to organize and maintain it by volunteering, mobilizing and 'participating'.

Table 10.2, whose framework is inspired by the work of Le Grand and Robinson (1984), summarizes the various ways that the State is involved in childcare services in the province of Québec (to simplify presentation, no distinction has been made between the federal, provincial and municipal roles). There are virtually no state-run pre-school daycare centres (cf. Prentice 1988 on the municipal role in daycare in Ontario) and, in any case, state provision would not mean free services. Subsidies to certain types of organized centres and to parents are, nonetheless, important. Moreover, the fact that parents may also receive tax deductions when 'informal' services of paid caregivers are used means that indirectly the State is also involved in sanctioning the existence of informal care.[5]

The State is heavily involved in regulation of all organized services to ensure that certain norms of safety, salubrity and, in some cases, pedagogy, are met. Regulation standardizes and rigidifies these services by controlling, for example, maximum opening hours, ages of children cared for and conditions under which daycare spaces are subsidized. For example, in Québec, some pre-school centres accept young children of school age in the summer or on school professional days, but no operating grants are available for daycare spaces used in this way. In Québec, family home caregivers ('babysitters') who do not operate as an official business and who care for only a few children at a time do not have to be licensed.

With these conceptual dimensions in mind, I shall now explore the origins of the current childcare situation in Québec and consider how

191

Table 10.2 State intervention in childcare provision in the Province of Québec

---

STATE PROVISION

Free services

– Pre-school group daycare and before and after school group care: free services
 are non-existent
– Kindergarten (age 5), usually ½ day ⎫ non-compulsory part of public
– Pre-kindergarten (age 4), usually ⎬ education system. Not considered
 2 hours/day ⎭ to be 'childcare'

Subsidized services

– Municipally-run pre-school daycare centres (very rare in Québec)
– School daycare services (set up in response to parent pressure)

STATE SUBSIDY

Subsidy to services (start-up and operating costs)

– Parent-controlled licensed pre-school daycare centres
– Licensed family home daycare services
– School daycare services

Subsidy to individuals (parents)

– Direct financial aid to very low income users of licensed services (pre-school
 and school)
– Tax deductions and other fiscal measures to offset costs of any mode of
 childcare (for under-12s) outside nuclear family (receipts required)

STATE REGULATION

Obligatory licensing of services

– All types of pre-school group daycare centre (commercial and non-profit)
– Family home daycare services (beyond a threshold number of children)
– School daycare services

Obligatory licensing of personnel

– All types of pre-school group daycare centres
– School daycare services

this historical legacy influences the current accessibility of services to different groups.

MODES OF CHILDCARE PROVISION IN QUÉBEC: POLICY DEVELOPMENT AND DETERMINANTS OF ACCESS

Historically, in Canada – not unlike the position in the United States (see Grubb and Lazerson 1982; Joffe 1983; Mann 1986) and Britain (see Dale and Foster 1986: 110–11; Lewis 1984: *passim*; Pascall 1986: 72–84; Ruggie 1984) – the notion of child*care*, seen as a social service, has been generally separate from that of child *education* ever since the introduction of compulsory primary schooling in the late nineteenth century (Canada, National Council on the Status of Women 1986: 230–4; Lero and Kyle 1989; Pence 1987; see also Table 10.2). This separation – which does not exist to nearly the same extent in many European countries which have extensive networks of pre-schools for children aged three and up (see International Labour Organization 1988; Sorrentino 1990) – entrenches the socio-economically and politically determined boundary between family and state responsibilities. It thus tends to increase the difficulties faced by parents with children who have to be 'educated' through the school system for part of their parents' paid working day and 'cared for' through childcare services for the remainder of the day, and for those at the point in the life course where they have some young children needing 'care' and others needing 'schooling'.

State assistance for daycare centres has been based on 'welfarist' principles since the early twentieth century, with centres being set up occasionally for children from mother-led families where the 'normal' model of full-time care by the mother could not be followed because she needed to take employment outside the home. Daycare facilities were also temporarily expanded during the world wars, when young married women's employment was considered both legitimate and essential to the economy. By contrast, state nursery schools and kindergartens were a non-compulsory wing of the education system, expanded since the 1960s as they increasingly came to be seen as a cultural, pedagogical and even economic tool for improving the opportunities of children from deprived backgrounds. Thus for instance in Québec about half of all five-year-olds attend state school kindergarten (Québec, Ministère de l'Éducation 1989) but only a few per cent of four-year-olds have access to pre-kindergarten, which is offered only in inner-city schools. Open for very limited hours (now four hours per day in Québec for kindergarten and two hours for pre-kindergarten, for example), these services were never intended as a response to the childcare needs of employed mothers, although today many parents regard them in this way, using daycare centres or informal modes of care for the rest of the day.[6] The Canadian system means, moreover, that full-time daycare expenses are often incurred for as long as five years per child, since full-time schooling is not compulsory in Canada until the age of six.

Similarly, the State has not accepted any responsibility for the care of young schoolchildren outside school hours; the school system thus effectively structures parents' (especially mothers') schedules and the strategies they adopt for combining the requirements of the school day, domestic labour, childcare and employment (Smith and Griffith 1989).

Spectacular increases in the labour force participation of mothers with young children from the mid-1960s onwards rendered increasingly untenable the traditional view that group daycare was only needed under exceptional circumstances. By 1989, mothers were in the labour force in 65 per cent of two-parent families with pre-school children (up from 32 per cent in 1976); two-thirds of these were employed full-time (Statistics Canada 1989a; 1989b). Half of Canada's two-earner families would fall into poverty without wives' paid work (Canada, National Council on Welfare 1988; Moore 1989). Moreover, mother-led families are no longer viewed as unusual: census data show that in 1986 about one in six families with dependent children was headed by a female lone parent.

Faced with the steady growth in mothers' labour force participation rates, and mounting political pressures from childcare activists and women's groups – later joined by associations representing commercial daycare (Prentice 1989: 19) – the Canadian government, much like its Australian counterpart (Franzway, Court and Connell 1989: 70–2), enacted legislation in 1966 that facilitated the development of cost-sharing arrangements with the provinces for daycare-related expenditures, through the Canada Assistance Plan; this was extended in the early 1970s (Mackenzie and Truelove 1991). In keeping with the liberal orientation of the Canadian welfare state (see Esping-Andersen 1989: 25–7; O'Connor 1989), daycare was not, however, to be free and universally available; nor (with a very few exceptions) were the federal and provincial governments to set up and run services. In the childcare field, as in other areas of Canadian social policy, '"by comparative standards [state expenditures have] represented a relatively limited response to the social tensions inherent in a capitalist economy"' (Banting 1987: 315, quoted in Ontario, Ministry of Community and Social Services 1988: 523).[7]

Within the broad parameters of the federal legislation, provincial governments could subsidize voluntary bodies (usually private, non-profit) and even commercial agencies to set up and maintain pre-school all-day daycare centres (see, for example, Bracken, Hudson and

194

Selinger 1988). They could also subsidize the fees in organized daycare services for those parents deemed to be the 'most needy'. Paralleling these measures, the federal government introduced income tax allowances for expenses incurred for any kind of childcare for children under twelve years (except that by a parent or minor sibling).

The Québec state, however, remained hostile to the employment of married women until well into the 1960s, as this was seen as a threat to the transmission of Franco-Québec culture through the family. Even with the increased demand for female labour and the general modernization of Québec culture during the 'Quiet Revolution' of the 1960s to early 1970s, organized childcare services received little public support. Starting in the late 1960s, impetus for change came from community groups working in impoverished inner-city areas of Montréal to develop services supportive of the local population and run democratically by their users (Godbout and Collin 1977; Granger 1987). Only after some of these groups had set up some pre-school daycare centres with the help of federal job creation money (on a tenuous financial basis and using mainly volunteer labour) did the provincial government feel forced in 1974 to use the provisions of the Canada Assistance Plan[8] to begin subsidizing places and providing direct financial aid for low-income families needing daycare.

The emerging feminist movement was an important element in the diverse coalition of interests that elected the nationalist Parti Québécois (PQ) in 1976, and helped give the daycare issue visibility during the PQ's first mandate. The number of daycare places increased rapidly after 1976, although by nowhere near enough to meet even the most conservative estimates of need – once in power the PQ was cautious on the daycare question, since it was also keen to maintain its support from more traditional wings of the electorate concerned with the upholding of 'family values' (Maroney 1988: 28). Nevertheless, Québec's policy was widely seen as being progressive because it favoured non-profit cooperative daycare centres with a parent-controlled management board over other types of centre; only this type of centre receives an operating subsidy and substantial start-up funds.[9] Three-quarters of all pre-school daycare centres were of this type in 1988,[10] thus taking the notion of group daycare as an essentially *voluntary* service, albeit with some state support, further than in any other province.[11]

Also, Québec was the first province to pass legislation (in 1979, amended in 1982; see Kuiken 1986) to allow before- and after-school daycare services to be set up by school boards with start-up and operating subsidies from the Ministry of Education if parents could show that there was sufficient demand in a particular school. As a result, the number of places doubled between 1982 and 1987 (Association des services de garde en milieu scolaire du Québec, 1989; *Petit-à-Petit* 1988), although density of provision varies considerably, geographically.

Although the number of licensed daycare spaces continued to increase in Québec throughout the 1980s – most of the increase in the second part of the decade being in school daycare services – in 1988, places in daycare centres were still only available for about 18 per cent of pre-schoolers whose mothers were in the labour force, with licensed home daycare services covering about another 2 per cent. As for young school-aged children (aged six to eleven), only one-tenth of those with employed mothers had access to before- and after-school daycare services (calculated from data in Québec, Ministère du Conseil exécutif 1988); the low figure reflects the disregard for childcare needs in this part of the life course until recently.[12]

Childcare has nonetheless moved from the margins to the political mainstream (Prentice 1988). Women's paid work is increasingly recognized as essential to the economy, and the desire to cultivate and retain the valuable pool of female professional, managerial and skilled technical labour is growing.[13] Childcare is seen as a key to this – but not one that will be given out freely.

I shall now take a more detailed look at how childcare-related policies affect the accessibility of different modes of care to different groups of potential users. This part of the discussion is based on existing literature and the findings of a survey of 475 families with five-year-olds in half-day public kindergarten and/or six- to seven-year-olds in Grade 1 (i.e. their first year of compulsory full-time education), conducted in 1987–8 in three Montréal districts, heterogeneous in terms of socio-economic position, family structure and ethnicity.[14]

THE COST OF CHILDCARE TO PARENTS

Group daycare is expensive. In 1988 fees in Montréal were close to C$100 per week per child in pre-school care, and about C$50 for before- and after-school care. Only 17 per cent of those using pre-school daycare centres received direct financial aid in 1988 (Québec, Conseil du statut de la femme 1989) (data are not available for school daycare services). Moreover, examination of the sliding subsidy scale used in Québec at the time of our survey reveals, for example, that a couple with one pre-school child and combined earnings about equal to the poverty line would only have received a subsidy for 50 per cent of the daycare fees. With a pre-schooler and a young school-aged child and earnings slightly above the poverty line, one-fifth of pre-tax earnings could easily have been spent on group care (after subsidy but before tax rebate).

The complementary relief available in 1988 through the federal and Québec income tax systems consisted of deductions amounting to

C\$4,000 per year per child in pre-school care and C\$2,000 for young school-aged children. For parents in higher tax brackets these deductions can reduce costs by as much as half, but they are worth much less to those on lower incomes. Moreover, rebates come only after tax returns are filed. Also, parents dependent on 'babysitters' often have a great deal of difficulty in obtaining receipts, which have to be filed with the Québec income tax return for amounts over $600. A not atypical situation was that of a woman we interviewed who relied on a sitter to care for her two-year-old all day and her six-year-old before and after school; this cost C\$75 per week, but the sitter would not give receipts for fear of having her welfare cheque reduced. The formal system would have been cheaper for this mother as she would have received a subsidy and a much bigger tax deduction, but accessible daycare services did not open early enough for her to be at work at 7.00 a.m.

FAMILY AND HOUSEHOLD STRUCTURES

While cost remains the key determinant of access to different childcare modes, other factors are also important, including family structures and household composition. In keeping with the still-dominant view that daycare centres are first and foremost a way of ending welfare dependency among lone mothers, parental subsidy scales and tax deduction rules somewhat favour single-parent families over two-parent families at an equivalent level of income (Canada, National Council on the Status of Women 1986; Québec, Ministère du Conseil exécutif 1988). As regards pre-school daycare centres, all available Canada-wide and Québec data indicate that the financial aid structure has created two major groups of users: single parents with low-to-modest incomes, and fairly affluent two-parent families, while virtually excluding two-parent families with low-to-modest incomes. As for school daycare services, our survey findings pointed to a broadly similar type of polarization of users; in the inner-city school daycare services, however, affordability problems were mitigated by the existence of various additional operating subsidies which reduced costs for all users.

It is thus not surprising that our survey findings indicated very heavy use of informal modes of childcare among two-parent families with blue-collar or service jobs. Here, family life history and the changing character of immigration to Canada played significant roles in shaping childcare options. Where extended family members were present, they were often relied on heavily for childcare. Although recent immigrant families were often interested in organized daycare services because of the opportunities for language training and socialization of children into a new culture, frequently they could not afford these as they were only eligible for a very small subsidy, despite modest family earnings. Care by extended family members cost much less and was sometimes offered free. Other recent immigrants in our survey (some of whom were political refugees) had no extended family members in Montréal; rather than use an unrelated caregiver, the two parents – typically both service workers – would arrange childcare between themselves by working different shifts (see Allen 1979); this is clearly the cheapest solution, though one that may have considerable costs in terms of family life if there is little overlap in the shifts worked by each parent (see Christian 1990). In other cases, compromise arrangements are possible; one couple we interviewed covered childcare in the before-school period by working partially overlapping shifts, but used the school daycare service, which was inexpensive in the inner-city school their five-year-old attended, in the afternoon when kindergarten class ended. Their ten-year-old looked after himself after school.

Ethnic minority lone mothers in our survey, most of whom were originally from Haiti or the English-speaking Caribbean, also depended heavily on unrelated caregivers when they did not have female kin in Montréal. One mother we interviewed, a hospital worker, opted to bring her sister to Montréal from Barbados to help out with childcare for a year when both her children were pre-schoolers, rather than relying on 'babysitters'.

Our interview findings also suggest that among families who immigrated prior to the mid-1970s – especially those from Italy, Greece or Portugal – a grandmother or other female relative was often the caregiver of choice, regardless of cost considerations, great emphasis being placed on the role of the older relative in passing on the language and key elements of the culture of origin to the young children. In these groups, such women had also reached the age of retirement from work in the formal sector. One grandmother we interviewed – recently retired from a Montréal clothing factory where she had worked since coming to Canada from Portugal eighteen years previously – had an eight-year-old grandchild living with her and also collected another after school and cared for him until one of his parents came home; this arrangement was straightforward, as all three generations lived in the same building. Such co-residence of the grandmother, in the same household or in an apartment in the same building is a common pattern among Montréalers of southern European origin.[15] Typically, entire families emigrated in the 1960s and early 1970s under immigration policies which favoured family reunification and hence contributed to the 'availability' of elderly relatives. This contrasts with the situation of Caribbean women who often originally came to Canada alone under special immigration programmes designed to create a labour supply of domestic workers (Seward and McDade 1988); after two years they could apply for permanent residence but their extended family networks were often left behind in their countries of origin (Meintel *et al.* 1985).

YOUNG FAMILY LIFE COURSES AND THE AGE GROUPS OF CHILDREN

The age groups of children also make a difference to how childcare can be organized; this of course changes as families move through the life course. Among other things, arrangements will be altered and financial or logistical problems may be resolved while new ones may be created as children move from one pedagogical/administrative age category to another. Moreover, the State can make arbitrary changes from one year to another in defining these categories and the services and costs associated with them. For instance, the upper age limit for eligibility for the maximum C$4,000 tax deduction for childcare expenses was increased from six to seven in 1988 (Québec, Ministère du Conseil exécutif 1988: 54). To take another case, logistical problems can arise when families reach the point in the life course where a child living within a specified distance from the school is considered too old to be picked up by the school bus. This was about to happen to a family we interviewed whose children were aged eight and four; the school bus pick-up was crucial to their childcare and job-scheduling arrangements and they were afraid they would have to move because of this administrative rule. Here we can see instances of how the material significance of the point one's children have reached in the life course is substantially shaped by the State; once again, this underlines that it is erroneous to conceptualize the life course as a purely natural, biological phenomenon (see Finch 1989: 136–7).

Parents with children in both pre-school and young school-age groups often have greater logistical and/or financial problems than those with only one child, or with two who fall into the same category of childcare needs; if they have a third child in the half-day kindergarten category, the situation is even more complicated as parents resort to multiple and patchwork strategies of childcare. Again, in two-parent families, working split shifts may be a way around the difficulties – an option not of course open to lone mothers without access to convenient formal daycare and school daycare. If, however, the family also contains children old enough to babysit or take a younger sibling to and from school, arrangements can be worked out more easily, as was often the case among the more settled and financially stable families of southern European origin we interviewed, whose parents were older and families larger. Reaching the life course point where the family included a mix of older and younger children sometimes alleviated things for lower income families, as well. In the case of one mother of Caribbean origin, both older siblings and an extended family member were marshalled into a childcare strategy that, on the whole, worked well. Her sister lived permanently with the family and was only employed outside the home for three hours in the morning. Between this relative and the two children in their early teens, care arrangements could be made at home for the nine-year-old before and after school, and for the youngest child outside of kindergarten hours. Accompanying the younger children to school and kindergarten was not a problem, either. One gap existed in their childcare schedule, however: the kindergarten class was only two hours long and the child had to wait over an hour afterwards to be collected. This mother, a nurse's aide, had always relied on her sister or a family friend, even when all the children were younger, because the workplace daycare 'is too expensive – it's for the doctors and nurses'.

LOCATION

Location relative to homes, workplaces or schools is an ever-present factor in access to different modes of childcare. Indeed, the micro-geography of childcare arrangements is crucial for understanding questions of potential supply and parental options, because of its implications for parents' time–space constraints and thence for scheduling the working day. Even small distances between home and school can become almost insurmountable problems if, for example, one has to leave for one's job before school daycare opens and there is no one to take the child to the daycare service. The issue of location intersects with and forms the material backdrop to some of the other factors discussed above. For instance, using the extended family for regular care is critically dependent on the relatives in question living close by. In other cases, mothers relied on informal networks because there was very little formal provision available near enough to where they lived or worked to be fitted into their employment schedules. Yet even access to 'babysitters' depends on living in an area with a supply of women available to babysit (see Wallman 1984: 117n), and in major cities this is increasingly rare, except in areas where many women are on welfare or other fixed income benefits and therefore do not want to give receipts.

Location can complicate daycare arrangements even more when children are in different age categories. The institutional and ideological separation of childcare from child education noted earlier usually entails a spatial separation as well. The three-year-old child of one recently-immigrated family we interviewed attended a daycare centre located 3 km away from the school his sisters (aged eight and nine) attended. In this case the father got home from his job at 3.30 a.m., awoke at 7.00 a.m. to drive the youngest to daycare and his wife to work, and returned home to see the other two children off to the nearby school. They came home for lunch, which he made for them; then he would collect the little boy from daycare at 2.50 p.m. His wife came home just before he had to leave for work at 4.00 p.m. After a 50 per cent subsidy, daycare cost this couple C$50 per week and the parents found this worthwhile, in spite of the travelling. Clearly, however, this family's life would have been easier had the daycare been located at the older children's school – yet such spatial arrangements are very much the exception rather than the rule. In suburbs where the population is rapidly growing, the neighbourhood may well have been planned along traditional lines and centred on the local elementary school, but rarely is there space to accommodate a school daycare service or a pre-school daycare centre there. In inner-city areas pre-school daycare centres often occupy (at a low rent) unused space in schools where enrolment has declined, but no mechanisms exist to give priority to the siblings of children attending the school.

More generally, with the obvious exception of employer-sponsored daycare, there is no clearly-defined state policy of promoting the location of non-profit daycare services in areas of particular need (except in rural areas); moreover, private market rents often have to be paid, and thus financial restraints allow little room to choose the best location.[16] With respect to school daycare services, school boards are supposed to promote provision in inner-city areas (those with high levels of deprivation, as measured by the techniques of factorial ecology). In areas undergoing gentrification this designation can benefit upper-middle income groups as well as low-income populations needing school daycare, while those living in suburban apartment districts, which are often 'pockets of poverty' surrounded by more affluent zones, may pay more.

The role of location, and its intersection with labour market situation, can be underscored by comparing the situation of two single parents we interviewed, living in very different kinds of neighbourhoods. The problems of the single parent mentioned earlier, who had to pay C$75 a week to a caregiver who would not give receipts, in large measure reflected the absence of daycare services in the area open early enough for her to take the children and still get to work at 7.00 a.m. Like many black women from the Caribbean, she had found that the best job available was at the hospital. Her neighbourhood, a post-war inner suburb, has become Montréal's newest 'reception area' for thousands of refugees and other recent immigrants from south-east and south Asia, Latin America and sub-Saharan Africa (Bernèche 1990; Rose and Chicoine 1991). These are young families who know little of the new culture and the majority languages, receive inadequate support from overstretched local social agencies and have great difficulty making ends meet. Eventually this mother found a daycare centre with its own transportation service which picked up her younger son around 6.00 in the morning, but the centre was several kilometres away, which was a worry for her. There was no school daycare for her older son.[17]

In sharp contrast was the situation of a French-Canadian single mother of two children aged eleven and six, who shared childcare with her ex-husband. She lived in an inner-city area with a high density of formal childcare services due to its long history of community organizing, a few minutes' walk from school and from her workplace as an office worker. Both children used the school daycare service before and after school, for which she got a substantial subsidy and claimed tax reductions. The earnings of the two women were similar, and both were well-informed about 'the system', but the latter had far better access to formal services due essentially to her 'standard' working hours and proximity to a school offering a daycare service. Here we can see some effects of the intersection of location and ethnicity – for ethnic minority women are more likely than white women to work non-standard hours (Martin 1984; Phillips and Moss 1989; Seward and McDade 1988: 37).

RIGIDITY AND FLEXIBILITY IN CHILDCARE PROVISION

This brings us to another key issue. As mentioned earlier, the rigidity or flexibility of service delivery may be a very important determinant of access; as far as the question of schedules and hours is concerned, this dimension is conterminous in the current childcare 'system' with that of 'formal' and 'informal' (see Table 10.1). The formal system was not set up for the needs of parents who work irregular or non-standard hours (Christian 1990; Friendly, Cleveland and Willis 1989) – and we have already given examples of the types of difficulties this can create in arranging childcare, particularly in families which include pre-schoolers and young school-aged children but no older children or co-resident kin. In theory, a better availability and coordination of transportation arrangements between informal caregivers, daycare centres and schools could help deal with some of these problems; in practice, a complex web of bureaucratic and financial obstacles usually strangles the possibility of action.[18]

The hours of licensed home daycare services can be somewhat more flexible than those of group daycare centres and school daycare services, while informal services whose very existence is negotiated on an individual basis between parent and provider are obviously the most flexible of all. It is thus not surprising that informal modes are usually selected by mothers whose schedules are irregular, especially when a relative is not 'on hand'. The ultimate in flexibility is the situation where the two parents arrange to work different shifts so that one is always available when childcare is needed.

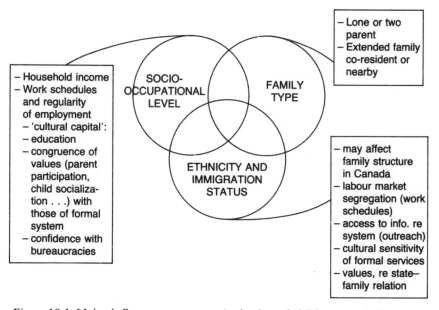

Figure 10.1 Major influences on parents' selection of childcare mode for young school-aged children

DETERMINANTS OF ACCESS: A SUMMARY

Based on the preceding discussion, it is possible to summarize some of the major influences on how Montréal parents select childcare services. Figure 10.1 provides a résumé of the situation. As can be seen, parents' 'socio-occupational' level (an awkward term, but encompassing more than 'class' in a narrow sense) intersects with issues linked to ethnicity and immigration status, and with family structure.

Broadly speaking, our results show, even after controlling for location in an area with greater or less formal provision, that Canadian-born families where the parent(s) had white-collar employment used the formal system most (for details, see Chicoine, Germain and Rose forthcoming; Rose and Chicoine 1991). Those with white-collar jobs tended, moreover, to be much more able to deal with the complexities of the financial aid provisions than those immigrants who worked in low-level services or manufacturing and were less able to appropriate 'cultural capital' (see Bourdieu 1984).[19]

For those not using the formal system, access to extended family members for childcare seemed to be a crucial dimension in how well childcare needs were met. According to our survey, longer-established families of southern European origin, whatever their income levels, seemed to manage particularly well because they usually had recourse

203

to locally-resident kin, who provided a reliable and inexpensive form of childcare while passing on key aspects of the culture of origin. By and large, these families appeared not to be interested in formal services.

Here, though, it is important to add some caveats. The putatively ubiquitous availability of the extended family in immigrant communities, and its cultural role, have long been used as justifications for failure to address the particular needs of such communities through the formal community services system (Rose and Chicoine 1991). But, as Taylor (1988: 124) points out, this type of attitude 'beg[s] the question about whether people might prefer to use formal services if they were delivered in a different way'. It is possible, for instance, that if organized daycare services were easier to set up, cheaper in all areas and sensitive to the twin desires of many immigrant parents to socialize their children into the majority culture and language while maintaining key values from the culture of origin, that more parents would opt for such services on a regular basis, even if they continued to rely on kin in times of illness or crisis (see Finch 1989: 54). And, furthermore, 'whether the different patterns of support which we can see in minority ethnic kin groups . . . persist as patterns of settlement stabilize' (Finch 1989: 52) is an empirical question defying generalizations across or between ethnic groups. In middle-class North American society, for example, retired people increasingly have the choice of opting for 'a more independent lifestyle, [which] can be incompatible with childcare commitments' (Wash and Brand 1990: 21). Moreover, as labour force participation rates increase for older women over the coming years, their availability for childcare will depend in part upon the age at which they become grandmothers relative to when they retire from paid work.

Furthermore, consideration is rarely given to whether, for instance, 'grandma' still needs to be employed in the formal work force or would prefer to attend a language class instead. Thus, regarding the extended family as a substitute for state funding in a context of patriarchal family power structures tends to 'naturalize' and reinforce the notion that the first recourse should legitimately be to older women when no parent is available for caregiving, because they will always put their children's and grandchildren's needs above their own. Life course stage is not in and of itself, then, a determinant of the availability of elderly women as childcare-givers.

In many cases this resource is simply not available. If anything, those families already in the most precarious of straits are also those least likely to have relatives at hand in their new country. In our survey those in the most difficult situations included those ethnic minority single mothers who not only occupied low-level service or manufacturing positions in a labour market segmented by gender and ethnicity, but also did not have ready access to extended family networks to help with

204

childcare. Recourse to unrelated caregivers was particularly heavy among this group, and while some were satisfied with the flexibility afforded by this type of informal market service, instability of supply and the question of receipts were perennial problems. Moreover, these respondents frequently used a number of different childcare arrangements in the course of a week – and recourse to such multiple strategies generally creates stresses for parents and children (see Allen 1979).

Recent immigrant couples (often refugees) without an extended family, who remain in low-wage 'ghettos' in order to manage childcare themselves through split shifts, are also in a particularly invidious situation, since usually they earn between them just enough to reduce substantially the financial aid they can get for daycare. Moreover refugee claimants waiting for official status (which can take years) are ineligible for childcare subsidies. In other cases, the mother stays at home because the parents cannot find two jobs on different shifts.

More generally, the cohort of Third World women who arrived in Québec since the economic downturn and restructuring started in the late 1970s is much more likely to be unemployed, is more concentrated in low-level jobs and has much lower average earnings from full-time work than earlier arrivals.[20] Moreover, a gradual process of gentrification has made it difficult for new immigrants to settle in the inner-city areas that now have the best service provision as a result of the legacy of community organizing in the 1970s and the more recent mobilization of middle-class activist parents; and poverty combined with racial discrimination in housing have pushed many into under-resourced inner suburbs (Bernèche 1990) – which, as we have seen, further reduces their childcare options.

Our findings suggest that while it would be a dangerous exaggeration to speak of the 'gentrification of daycare',[21] it must be admitted that the cost structure, the emphasis on 'voluntarism' in the setting-up and maintaining of services and the locational bias this tends to bring with it, do generate tendencies in favour of white-collar workers and professionals. School daycare services seem unlikely to be established in areas without a strong 'new middle-class' element, unless much more is done to reach out to those not otherwise in a position to organize for services and help them cut through the daunting bureaucratic obstacles that lie between the expression of parent demand and the inauguration of a service (see Rose 1990 for a more detailed discussion). Moreover, unless daycare subsidies for low- to modest-income families are greatly increased, we may expect to see – as is already happening due to recent changes to the financing formula for non-profit centres in Québec – more and more empty places in pre-school daycare centres in working-class neighbourhoods, because local residents cannot afford them.

Furthermore, our research indicates that the logistics of arranging daycare are compounded when the family is at the point in the life course where there are two or more young children needing different types of care at different times of the day, and no older siblings or friends to help out. These problems are due not merely to the passage through a natural stage in the life course, but are mediated by class and ethnicity and exacerbated by the rigid categories created by the institutional/pedagogic system. Such difficulties are most onerous for families whose job schedules do not fit the 'standard' hours of the formal system, and despite the gains of the women's movement most of this burden still falls on women – especially immigrant women.

CONCLUSION

Our research findings underline the diversity of childcare needs among families with young school-aged and/or pre-school children in contemporary metropolitan areas. They chime with recent calls for a greater flexibility in childcare options, so that more parents might be able to find affordable and culturally-sensitive quality care in the right place at the right time, without forcing them into undesired types of jobs and awkward schedules (Canada, National Council on the Status of Women 1986; Canada, National Council on Welfare 1988; Phillips and Moss 1989; Québec, Comité de la consultation sur la politique familiale 1985; 1986).

These types of demands influenced, and were influenced by recent feminist thinking about 'humanizing' the welfare state (see, for example, Adamson, Briskin and McPhail 1988; Borchorst and Sim 1987; Croft 1986; Finch 1986; Franzway, Court and Connell 1989; Greater London Council 1986; Waërness 1989), and framed within an overall set of recommendations that called for a much *greater* state commitment to financing and helping groups organize diverse formal and informal services themselves (as in Denmark; see Phillips and Moss 1989). But in Canada as elsewhere (Johnson 1989: 21–2) the whole meaning of 'flexibility' is being subverted by the New Right. Legitimate critiques of the 'rigidity' of state-regulated services are being employed to justify cuts in aid to organized daycare. As a result, 'freedom to choose' could increasingly mean, for all except affluent parents, 'choosing' from a range of services offered by caregivers with low pay and insecure working conditions, rather than by fully-recognized providers of a community service (Myles 1988: 97; see also Christopherson 1989). If current trends continue, the onus of ensuring 'flexibility' may thus come to rest increasingly on these low-paid caregivers as well as on parents and other family members, as formal services become increasingly 'informalized' (due to higher staff turnover and greater reliance on volunteer labour) and informal services fail to get the support they need.

This chapter has examined the role that a particular array of modes of service provision may have in perpetuating socio-economic and ethno-cultural divisions among single- and two-parent families with young children. It has highlighted the situation of recently immigrated and ethnic minority women, not because it is different in *kind* from that of other low-income women at the same point in the life course, but because it is often exacerbated by discrimination, by concentration in jobs requiring non-standard hours and by even more difficulty in appropriating 'cultural capital' than that experienced by Canadian-born white working-class women. In addition, there is concern that, due to economic restructuring, the particularly severe difficulties faced by recent immigrants in establishing themselves in the labour market may prove to have a permanent effect on the economic situation of this cohort as they move through the life course. Thus, finding appropriate childcare arrangements may be even more crucial to these parents' abilities to find and keep their jobs (Fincher 1989; Hanson and Pratt 1988) than for other groups, especially if care by kin is not an option. Recent immigrants with neither French nor English as mother tongue now form a large and growing minority in Montréal area elementary schools; their particular childcare needs must thus be given more attention.

Conceptually, childcare needs to be resituated in the larger context of the relationship between economic restructuring and the reshaping of reproductive work (see McDowell 1990; Sassoon 1987), as mediated by state policies which set the outer parameters of this crucial aspect of everyday life, usually experienced at a microscale of time and space. Future research on young family life courses needs to contextualize the findings of empirical work using ethnographic research methods that address, for example, people's definitions of their needs at different points in the life course (see, for example, Yeandle 1984), by exploring the role of the State, capital and the various institutions of civil society in defining life course stages and transitions, and structuring their material significance in everyday life. This will enable us better to interpret how people select childcare, how their options are structured and what the potential barriers are to organizing around 'common interests' for a genuine diversity of affordable modes of childcare provision.

11

WOMEN'S TRAVEL PATTERNS AT VARIOUS STAGES OF THEIR LIVES

Sandra Rosenbloom

INTRODUCTION

There has been relatively little in-depth analysis of the relationship between the sets of responsibilities inherent in employment and child-care and domestic obligations, and women's transportation choices. Equally ignored have been the travel patterns of elderly women, particularly those who are alone. The few researchers who have focused on working mothers have generally concentrated on the home-to-work trip only; the available data have limited explorations of non-work activities and differences in travel patterns created by race or ethnicity, marital status, or variations in occupation. Elderly women's travel patterns have generally been ignored because the elderly are seen as a monolithic group, whose needs are assumed to be known, and because they rarely make the oft-studied work trip.

This chapter summarizes several studies which were designed to address some of these gaps in our knowledge of the travel behaviour of women at various stages in their life. For women in the labour force the focus was on the impact of marital status, the number and age of children, and household and childcare responsibilities on travel patterns. For elderly women the focus was the impact of having either a driver's licence or a spouse with a licence.

I first analyse the travel behaviour of full-time employed women, comparing single and married mothers, and evaluating the transportation implications of having children of different ages, examining the whole pattern of their daily trips – not just their work trips. The bulk of the material refers to women in the United States, but to gain some insights into the significance of cultural context in shaping women's travel patterns, I introduce comparative data from the Netherlands. Then I focus on elderly women in the United States, comparing single and married women, and contrasting those who do not drive – soon a small minority of that population – with those who do.

Both analyses show that traditional economic variables – which ignore crucial cultural roles and the salience of the life course – are inadequate to explain differences in the travel patterns and needs of women in various stages of their lives. The real mobility problems facing employed married women result from their complicated responsibilities and from the travel needs of their children; those of the single women from dealing with these complex responsibilities without a resident partner. The mobility problems of the elderly arise from their lack of alternatives when they can no longer drive or easily obtain car rides because they have based their housing and lifestyle choices on the access offered by the car. Without understanding how life cycle and related constraints affect travel behaviour, it is difficult to respond rationally to the needs of the variety of women in society.

The policy implications of these findings are enormous; traditional responses are questionable. Income transfer programmes, which tacitly assume that the plight of working mothers can be addressed only by income enhancement, will rarely be successful unless they are part of a package of solutions which addresses all the problems faced by working women. Public transport subsidies and extensions of traditional public transport services will be successful in only a few select situations, because public transport is rarely available or effective in the suburbs, where over two-thirds of most women work or live; even where available and safe, public transport imposes a terrible time penalty on working women. Planners and policymakers must understand how women organize their transportation resources to meet the myriad demands of their daily lives in order to address the mobility needs of women of any age.

BACKGROUND

The women whose transportation choices I discuss – working mothers and elderly women alone – constitute major segments of the female population. In the United States in 1985, 53.4 per cent of married women with children aged five and under, and 48 per cent of married women with children under one, were in the paid labour force. Over 75 per cent of the single mothers who head 20 per cent of all US families had salaried employment. Almost three-quarters of all employed women had full-time salaried employment. Almost 60 per cent of all people aged over sixty-five, and almost two-thirds of those over seventy-five, are women (US Census 1988: Table 13). The largest component of the elderly are women alone; currently more women over sixty-five live alone than live with spouses or relatives. With the average age of widowhood at 64.5 years, it is not surprising that so many older women are alone, or that elderly women are the fastest-growing element of society.

WOMEN'S TRAVEL BEHAVIOUR: THE LITERATURE

We know that employed women work closer to home, travel for a shorter time and distance to work, and more often use public transport than men. Little analytical research has focused on journeys other than the work trip, however, and few studies have gone beyond traditional economic variables at the household level – size, income, employment status – to explain these differences. Many researchers have explicitly rejected a major role in transportation decision-making for non-economic or individual variables such as race, ethnicity, cultural characteristics, or age.[1]

Yet a growing body of data testifies to the validity and need for investigating both non-work trips and non-economic variables. Early pioneers like Hanson and Hanson (1980) found that Swedish working women made more shopping and domestic trips than their spouses – and fewer trips for social and recreational travel. Moreover, the men were more likely to take the car in one-car households. Work by Rosenbloom in the Netherlands, France, and the United States found that women's travel patterns varied significantly with the age of their youngest child (Rosenbloom and Raux 1985). Perez-Cerezo (1986), in a very small study, also found that the age and presence of children influenced travel patterns in all types of household. A 1990 study in four Chicago suburbs found that employed women made *twice* as many trips as comparable men for errands, groceries, shopping, and chauffeuring children (Prevedouros and Schofer 1991).

Pickup (1985), studying British women in Reading, found that those with the greatest childcare obligations made the shortest work trips, passing up better jobs with longer commuting times. He concluded that women do not travel further because their childcare obligations – and not the travel costs – limit them. In support, he found that a significant number of women without children were willing to drive considerable distances even for low pay.

The role of marital status, too, has been ignored; researchers are often stymied by data sets which did not record marital status.[2] However, the few researchers addressing unmarried parents find the lack of attention to single mothers to be a serious omission. Kostyniuk *et al.* (1989) found that, except for the very poorest women, who did not drive, single parents made more trips and travelled further for all purposes than comparable married workers; they attributed these patterns to the need to balance employment and domestic responsibilities without the help of a resident partner. Johnson-Anumonowo (1989) found that although single women in Worcester, Massachusetts were less likely to own cars, they were more likely to make their work trips in cars; she also found that single women had longer work trips than comparable married women. She concluded that other variables – such as occupational differences, and differences in workplace and residential location – might explain these patterns and must be studied. Rutherford and Wekerle (1989) studied single and married workers in a Toronto suburb and concluded that it was important to disaggregate women by family composition; they found that single women spent more time travelling to work, and that they were less likely to work in the suburb in which they lived than were comparable married women.

Although racial and ethnic variables have not traditionally been considered in transportation research (observed differences being thought to represent only economic disparities) a few studies suggest the seriousness of this omission. Analysis of 1980 US census data on travel to work shows that Hispanics are more likely to carpool than comparable workers, and less likely to use public transport than others in comparable socio-economic groupings (Miller, Morrison and Vyas 1986). A 1982 study found that Mexican-Americans in Denver used public transport far less than comparably situated Anglos, preferring to share cars and travel with friends on all trips, and also because they were travelling to different places for their activities. McLafferty and Preston (1991) found that commuting times for Black and Hispanic women in New York City equalled those of Black and Hispanic men, but far exceeded those for both White men and women (although the authors conclude that this proves rather than negates the importance of simple economic variables in explaining travel differences, since race and income and industry are strongly related).

Issues of race and ethnicity overlap with concerns about the elderly, and particularly elderly women; Martin Wachs has noted the growing importance of such variables in studies of the elderly: 'these variables will play a larger role than age in identifying communities of the elderly during the coming decade' (1978: 13). Wachs and his associates found that variables such as ethnicity, race, and geographic location within a community significantly affected elderly transportation patterns. He established, for example, that elderly Mexican-American women in Los Angeles were significantly less likely to have a driver's licence but were more likely to make trips in a car than comparably situated Anglo or Black women.

An NSF study, also in Los Angeles, found significant differences among Black, Anglo, and Hispanic elders of comparable socio-economic status. For example, older Hispanics depended on their families far more often than in other racial or ethnic groups. Comparable elderly Blacks and Anglos, conversely, were more likely to drive themselves to meet their travel needs.

In sum, the traditional transportation literature has little to say about the trips that fill the days of most working women or their elderly mothers, nor about how differences in trip patterns might reflect

211

variations in personal preference (due to race or ethnicity or age) or in available household 'resources' (e.g. a resident spouse) or in domestic or childcare responsibilities. However, a growing body of work suggests that these cultural and life course factors do influence women's travel behaviour.

STUDY METHODOLOGY

This chapter is designed to address some of the key omissions of transportation research, by examining the daily travel of women at different points in their lives. The work focuses first on the impact of children on travel behaviour, particularly as they grow up, and then on the importance of marital status in explaining travel differences. Then the study turns to elderly women, with and without driver's licences, and with and without spouses. Both studies are based on a qualitative analysis, identifying discernible patterns of behaviour and relationships. Because the research is largely descriptive, it cannot prove that non-traditional or non-economic variables explain the travel behaviour of working mothers or elderly women while simple economic variables do not. Yet the patterns are so clear that preliminary observations, and their obvious policy implications, seem justified.

The analyses are based on three very different data sources. The analysis of single and married employed parents in the US is based on a small-scale survey of single heads of household and married parents in Austin, Texas. The analysis of Dutch workers is taken from a companion study to the Austin study, undertaken in Rotterdam in the Netherlands during 1986. The data on the travel patterns of elderly women alone is taken largely from the 1983 Nationwide Personal Transportation Study conducted for the US Department of Transportation by the US Census Bureau. Each of these data sources is briefly described below.

The first data source, the small-scale attitudinal studies, were undertaken by phone in 1982 (married respondents) and in 1985 (single parents) in Austin, Texas, based on a random quota sampling technique. Austin is a city in central Texas with a 1980 population of 484,000. Respondents were married and single women workers with children of various ages living at home. Both full-time and part-time workers were interviewed, but this chapter addresses only those working thirty-five hours per week or more (i.e. full-time). Note that all married men described here are married to full-time salaried women with children.

One hundred married and fifty single parent families were interviewed. These parents were asked a variety of descriptive, attitudinal, and hypothetical questions about their travel patterns and needs. The small scale of this data set makes it difficult to draw statistically significant conclusions, particularly when focusing on sub-groups within the overall study population. However, the data, while descriptive, can be indicative of larger patterns.

As a group the single mothers in the Austin study are very atypical; they have far higher incomes than the average single-parent family in the US or in Austin, and car ownership is almost universal. While the income issue was a concern to the researchers, it does mean that married parents are being compared to single parents with comparable household incomes.

A second comparable study was undertaken in two European countries in 1985–6: in Lyons, France, and in Rotterdam in the Netherlands.[3] Research procedures were roughly comparable to the Austin study, as was the instrument; all interviews were undertaken in the home rather than by phone. Only the Rotterdam survey is reported on here; that study interviewed the workers in 150 two-adult and fifty female-headed households.

The third data set is unpublished tape-readable data from the 1983 National Personal Transportation Study (NPTS), a nationwide survey of 6,438 households. Unfortunately these data were originally developed for another project[4] and do not always directly address the key concerns of this chapter; sometimes the analysis relies on proxies and indirect measures.

WOMEN IN THEIR WORKING YEARS WITH CHILDREN

This section focuses on women employed full-time outside the home as well as within it; it describes only women with children still living with them. The purpose of the study is to show how the age of their children and their marital status affects the travel patterns and needs of salaried women. Like both of the analyses in this chapter, the one below is forced to use cross-sectional data in an attempt to gauge longitudinal effects – such as the influence of growing children on their parent's travel patterns. It is wise to remember that the parents in the data sets below whose youngest child is a teenager are not the same parents whose youngest child was under six ten years ago.

GENERAL TRAVEL PATTERNS OF MARRIED AND SINGLE PARENTS

The following analysis examines the travel behaviour of men and women working more than thirty-five hours per week in terms of three variables: trips linked to and from work, trips made solely for children, and the way in which children themselves travel. The analyses show that all employed women with children have different travel patterns than comparable men, that these travel patterns change as the children in the family grow up, and finally, that single mothers have remarkably different travel patterns to either married women or men with children.

Linking work trips

Single and married parents were asked if they routinely or frequently combined trips for other purposes with trips to or from work; transportation planners generally describe such combinations as linked trips.

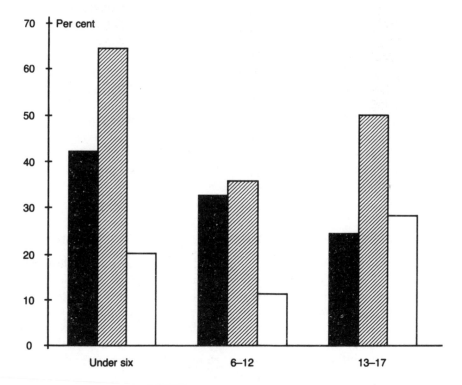

AGE OF YOUNGEST CHILD

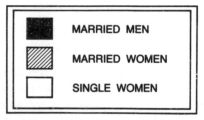

Figure 11.1 Percentage of parents who link trips to work

214

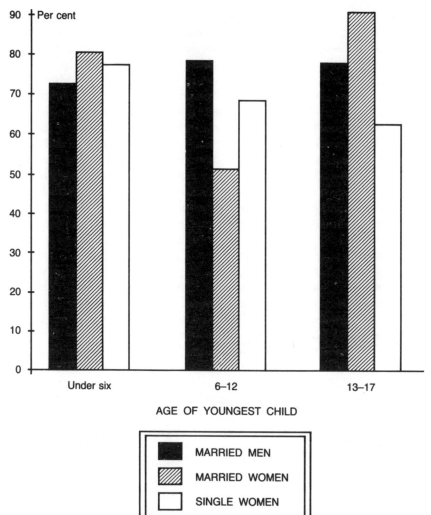

Figure 11.2 Percentage of parents who link trips from work

The existence of linked trips indicates complicated travel patterns which are not easily served by modes other than the private car. Figures 11.1 and 11.2 display parental responses; Figure 11.1 is concerned with the outward trip to work, while Figure 11.2 deals with trips home from work.

Figure 11.1 shows both the important differences between working women and men, and those between single and married female parents. First, it illustrates that men and women who had either very young children or teenagers had disparate travel patterns. Sixty-five per cent of all salaried married women with children under six linked trips to work; 42 per cent of comparably situated men did so. Working men were less and less likely to link trips to work as their children grew older.

Although the majority of married salaried women link trips to work when they have small children, less than a third do so when their youngest child is six to twelve, but half do so when they have teenagers.

Figure 11.1 also illustrates that single and married female parents have different linking behaviour to work, although almost all respondents drive.[5] Less than 30 per cent of single women link trips to work, whatever the age of their children; less than 20 per cent do so if they have children under twelve. Only with teenagers do single mothers combine more trips than either of the married parents: they are only slightly more likely than married men, but nowhere as likely as married women to trip-link.

Figure 11.2 shows that all parents are more likely to link trips on the way home from work. Here the linking behaviour of married men seems unaffected by the age of their children; roughly three-quarters of all men do so, regardless of the age of their youngest child. Slightly more married women link trips from work; while 80 per cent of women with very young children combine their trips, only half of married mothers with children aged six to twelve do so but almost 90 per cent of the mothers of teenagers link trips.

Single mothers of children under six act roughly the same as married parents; between 70 and 80 per cent of all such parents link trips on the way home. However, single mothers of children aged six to twelve are more likely to link trips than married women, but less likely than married men. Roughly 68 per cent of single women with school-age children link trips on the way home. This figure drops slightly for single mothers of teenagers.

These data can only hint at reasons for the differences between the sexes and between single and married women. It seems likely that married women have different patterns than their male partners because they accept more responsibility for their children's needs. Married women are more likely to make a greater variety of trips for young children, and more of those trips are directly related to household responsibilities (Rosenbloom 1989a). Single women may link trips less often for younger children because they are not as willing or – lacking a partner – able to leave their children for the time necessary to make linked trips. A married parent may be more willing or able to either pay for extra childcare or use the other parent to watch young children.

Trips made solely for children

All full-time working parents in the households surveyed were asked if they ever made trips solely for their children, and not because they

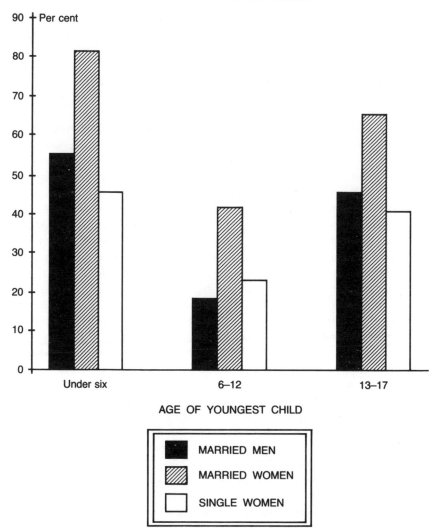

Figure 11.3 Percentage of parents who make trips solely for their children, by age of children

themselves needed to be somewhere;[6] transportation planners call these journeys serve-passenger trips. All three types of parents responded positively, although, as Figure 11.3 shows, the percentage dropped sharply as children aged, and there were important differences between married and single female parents.

Married women are the most likely to make trips for their children; almost 82 per cent of married women with younger children do so. Although the percentage of trips drops sharply as their children grow

217

up, almost half of all married women continue to make trips to chauffeur their teenagers. Slightly more than half of all men make trips solely for their children when the youngest is under six, but only 18 per cent of men do so for their teenagers.

With very young children, married mothers and fathers differ only slightly in the kinds of trips for which they drive their children; recreation trips are common for both parents, but fathers are more likely to take children to organized school or church activities. As children grow up, fathers, who are making few trips solely for their children, largely provide leisure trips. The small number of fathers (18 per cent) who chauffeur their older children make only recreation and social trips (Rosenbloom 1989a).

As children grow up, married mothers continue to provide a variety of trips, including medical and organized church or school activities. By the time their youngest children are teenagers, the 88 per cent of all women chauffeuring their children make *no* recreation trips but persist in providing trips to doctors, lessons, and organized school activities.

In all cases single women are only half as likely as married women to report making trips solely for their children of any age. Like most parents, the chauffeuring activities of single mothers decline sharply as their children grow up. Less than half of the single mothers of younger children report chauffeuring those children; the percentage drops to less than a quarter of single mothers of older children (only slightly higher than married men).

Single parents of teenagers resemble neither married parent when the trip purposes of the chauffeured trips are examined: single women are twice as likely to make recreational trips for their children as married women, but far less likely to make any medical or organized school or church trips (Rosenbloom 1989b).

Data on the frequency of chauffeured trips are illuminating (Figure 11.4); they show that most of the men who report making trips for their children do so far less frequently than married women. Of the 54 per cent of men who report making such trips for their children under six, the overwhelming majority, 75 per cent, do so less than once per week. Of the more than 90 per cent of all women with younger children who reported such chauffeuring duties, most did so once a week or more *per child*, with 16 per cent doing so more than three times per week per child.

Both parents are less likely to chauffeur slightly older children, although the drop as the children age is far larger for men; only 40 per cent of fathers but 64 per cent of married mothers make trips solely for children aged six to twelve. Of those, the majority of fathers make those trips once or twice a week. Mothers' patterns for these school-aged children are similar to fathers', although 15 per cent make trips more than twice a week.

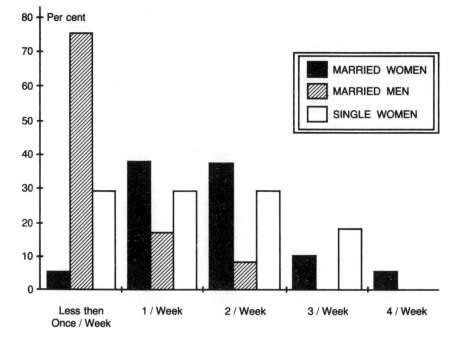

Figure 11.4 Frequency of trips made solely for children under six

Married mothers and fathers have very different frequency patterns for teenagers. The overwhelming majority of trips by the 18 per cent of fathers who make trips for teenagers occur less than once per week. Of the 46 per cent of married mothers who continue to make trips solely for their teenagers, over a third of such trips are made once a week or more, while 15 per cent are made three times a week or more.

While fewer single women chauffeur their children under six than either married parents, of those doing so, most do so more frequently than married men although generally less so than married women. Over 70 per cent of single women make one or more trips per week per child for those under six; over 40 per cent do so twice a week or more. While married parents individually tend to report less frequent chauffeuring of older children, single women show the reverse pattern. Single women make far more frequent trips solely for children aged six to twelve; almost a third of the 40 per cent of single mothers who chauffeur children of this age make such trips four or more times per week for each child.

These data show clear patterns, even if they clearly cannot show the underlying causes. Married mothers of very small children are over-whelmingly the chauffeur for these children; the majority of trips being made by married men for their children are made very infrequently and appear to serve a back-up or emergency function. Both mothers and fathers of children aged six to twelve make fewer trips solely for these children, although far more mothers still provide the trips needed. The frequency of trips does decline, so that the majority of both parents are making trips once a week or less for these children.

The drop in serve-passenger trips as children grow up may be caused by the fact that school-aged non-teenagers probably make fewer trips not covered by their school or caretakers, and they can walk alone to some destinations. Teenagers are probably more engaged in extra-curricular activities than younger children, but they may also have more alternative travel options than younger children. However, almost half of all married mothers feel obliged to continue to chauffeur teenagers. Few married fathers appear to be willing or able to involve themselves in teenage travel patterns; when they do so, it is very infrequently.

The data also show clear patterns for single mothers. As a group they are less likely to chauffeur their children than comparable married parents. We cannot tell if the children of these single parents travel less often or if they have other options available, options either not available to the children of married parents or which married parents are less likely to use. The next section addresses the issue of alternative modes for children.

Children's travel modes

Employed married and single parents were asked to describe their children's most frequent travel modes. As Figures 11.5 and 11.6 show, both married parents agree that the mother is the most frequent travel 'mode' for children of all ages. Roughly 60 per cent of all respondents report that the mother provides transportation most frequently. Interestingly, more married parents of both sexes report that 'other adults' are the primary travel provider than report that the father is. Roughly 5 per cent of all women and 2 per cent of all men report that the father is the primary travel mode for younger children; 7 per cent of all men but no women report that fathers are the primary provider of transportation for school-aged non-teenagers.

Equally-shared chauffeuring is more likely to be reported for younger children; approximately 12 per cent of both parents report that this responsibility is equally shared for children under six. However, no father or mother reports sharing this responsibility for older children.

Modes that do not require the presence of an adult, or more likely do not require a car, are not frequently reported. Approximately 11 per cent of married mothers report that their six- to twelve-year-olds relied more on bikes than on people.

Single mothers are slightly less likely than married mothers to be the

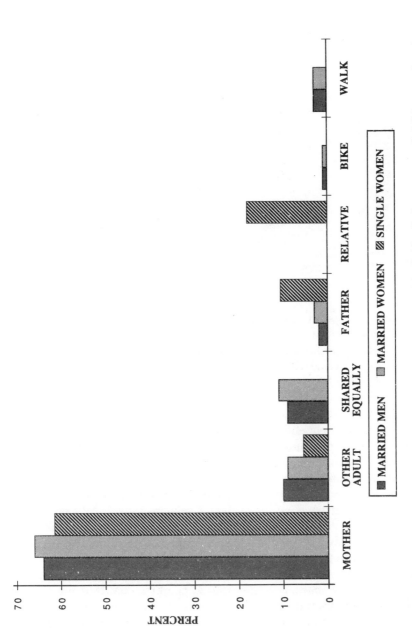

Figure 11.5 Children's most frequent travel mode as reported individually by parents of children under six

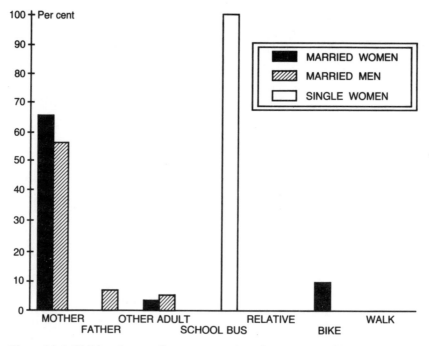

Figure 11.6 Children's most frequent travel mode as reported individually by parents for children aged six to twelve

primary transportation provider for younger children. Surprisingly, single mothers are more than twice as likely to report that the child's father is the primary provider of transportation – although presumably he does not live in the household! In addition, 17 per cent of single mothers of children under six report that relatives are the primary travel mode of these children, a response given by none of the married parents. Overall, approximately a third of all single mothers, regardless of the age of their children, report that related and unrelated adults have the primary responsibility for transporting their young children.

The patterns of single mothers are remarkably different from those of married parents for older school-aged children. Strikingly, every single mother reported that the school bus was the primary travel mode of her children aged six to twelve, a response given by none of the married parents. At the time of the survey, Austin was the scene of a recent massive court-ordered busing of children above the third grade. Perhaps married parents arranged other transportation for their children, while single mothers were unable or unwilling to do so.

Overall, these analyses suggest that some of the chauffeuring trips for children made by their married mothers are made by other adults for the children of single mothers. However, the data at least suggest that there is a 'gap' in chauffeured travel between the children of married and single parents. To fill the gap, the children of single mothers may make some trips independently and they may forgo other trips. Both situations can have far-reaching social consequences. We may have concerns about small children either left to travel alone or those who lack educational, social, and cultural opportunities because they cannot access these opportunities by walking or obtaining parental rides.

COMPARING THE DUTCH SITUATION

The section above shows that the transportation patterns of working women, whether married or not, are heavily influenced by their children's travel needs, as well as their other domestic responsibilities. Examining comparable data about Dutch working women and their children allows us to see how differences in the transportation resources available to Dutch children affect the travel patterns of their parents. Unfortunately, adequate data are available only for married Dutch parents.

Analyses of the Dutch data show that Dutch families as a unit are slightly less likely than their American counterparts to provide chauffeured travel for their children. The comparison is complicated, however, because Dutch mothers are as likely to provide serve-passenger trips for younger children and indeed more likely to do so for older children, than are American married mothers. However, most families do so less frequently than American families. Combining the greater likelihood with the lower frequency, it appears that Dutch families provide slightly fewer trips for children under six and far fewer trips for older children than American families.

However, for Dutch children of all ages, the parent providing the transportation is overwhelmingly the mother. So, Dutch married women are providing a sizeable number of serve-passenger trips – in spite of a full array of safe, efficient transportation alternatives for children – although not as many trips as comparable American women.

Trips solely for children

As in Austin, Dutch working parents were asked if they ever made trips solely for their children; the data in Figure 11.7 show many similarities and some important differences with American parents. Roughly 75 per cent of Dutch fathers of small children report making trips solely for their children – comparable to US data. However, not a single Dutch father of children over six ever reported making trips solely to take his children somewhere.

Dutch married women are as likely as American women to report

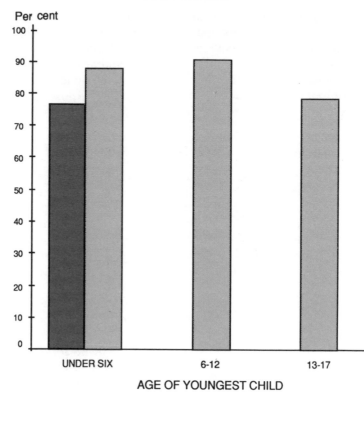

Per cent

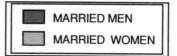

AGE OF YOUNGEST CHILD

MARRIED MEN
MARRIED WOMEN

Figure 11.7 Trips made solely for Dutch children of different ages by their parents

routinely making trips solely for very young children; they are actually more likely to report serve-passenger trips for six- to twelve-year-olds than American women. Overall, more than 70 per cent of all employed Dutch women report that they make trips only to take their children somewhere. It appears that Dutch women may be making up for the trips made by American fathers which are not made by Dutch fathers for children aged six to twelve.

As with American data, however, frequencies are illuminating. Figure 11.8 shows that Dutch women make far less frequent serve-passenger trips than do US women; over 60 per cent of all trips for younger

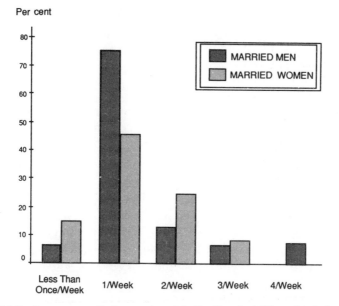

Figure 11.8a Frequency of trips made solely for Dutch children by their parents; children under six

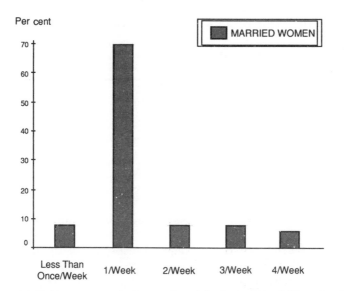

Figure 11.8b Frequency of trips made solely for Dutch children by their mothers; children aged six to twelve

children (under six) are made once a week or less. Close to 80 per cent of all serve-passenger trips made by Dutch mothers for children six to twelve are made once a week or less. In other words, although Dutch employed mothers are more likely to make serve-passenger trips than American mothers, the trips they make are made less frequently. So Dutch families as a unit are probably making slightly fewer trips overall for very young children.

However, the number of serve-passenger trips appears to be substantially less for Dutch families with children older than six. One possible explanation of the smaller role of Dutch parents in their children's travel is that the superior public and other transportation options available in Rotterdam free parents from involvement in all of their children's travel needs. The section below addresses this issue.

Children's travel modes

Dutch working women were asked if their children routinely or frequently travelled alone; the results are shown in Table 11.1; only 21 per cent of mothers reported that children under six travelled alone, but the number rose to 92 per cent for mothers of children aged six to twelve, and 100 per cent for teenagers' mothers. Table 11.1 also shows the travel choices of those children; they rely heavily on public transport and bikes (note that in the Netherlands, driving licences are not granted to those under twenty-one years old). Bicycles are, of course, a major feature of Dutch culture; they are also heavily used by adults and account for 25 per cent of the home-to-work commute in Rotterdam.

Table 11.1 Percentage of Dutch children travelling alone, and most frequent travel mode

	Per cent travelling alone		
	Under 6	6–12	13–17
	21	92	100
Mode of travel for those travelling alone (percentages)			
	Under 6	6–12	13–17
Public transport	58	58	43
Bike	8	42	43
Walk	34	–	9
Motorcycle	–	–	5

Clearly the existence of workable transportation options that are considered safe and secure enough for young children probably explains why Dutch women make fewer chauffeuring trips than their American counterparts. However, it is important to remember that travel options do not explain the differences alone – Rotterdam has the kind of dense land-use patterns that make these options effective. I have argued elsewhere that US land-use patterns are so diffuse that most American women cannot make the majority of their trips using any other mode than the car even if public transport services are available, given the location of their jobs and the significant time constraints imposed by even a hypothesized close-to-optimal public transport system (Rosenbloom with Lerner 1989; Rosenbloom 1991).

While existing transportation options help Dutch women deal with their children's travel needs, they are hardly a panacea. Even with extremely good public transport facilities and strong societal support for cycling, mothers still make many serve-passenger trips and their travel patterns are still strongly affected by their children's needs (Rosenbloom 1985a; 1985b). It is a fact that the availability of travel alternatives – whether these are used or not – seems to give Dutch fathers less reason to respond to their children's needs.

ELDERLY WOMEN

This part of the chapter focuses on the transportation problems of two types of older single women: first, those without driver's licences, with special emphasis on the 'stranded widow', or elderly woman who does not drive but depends on her licensed spouse until his death, and, second, the ageing female driver.

Today, between 30 and 40 per cent of women over sixty-five in the United States live alone and an additional 20 to 30 per cent do not live in their own households (US Census 1987). Although the percentage of women living alone has been dropping slightly for two decades[7] the figures are still in sharp contrast to those of comparable males; over three-quarters of men over sixty-five live with their spouse, and only 10 per cent do not live in their own household. Since the average age of widowhood is expected to rise to only 66.7 years by 2010, it seems likely that the single largest component of the elderly population will continue to be women living alone.

The analyses below show that the majority of elderly women in the future will be licensed drivers; as their skills deteriorate and their financial ability to maintain a car declines, their mobility loss may be significant. Elderly women drive less and make fewer trips than comparable men; while some of this difference in travel behaviour no doubt reflects a reduced desire for travel, some must be a reflection of the difficulty elderly women face in finding alternatives to driving when they can no longer do so.

These analyses, it should be cautioned, are based on cross-sectional data; although they tell us what women in different generations do, they cannot tell us what women in the same generation will do as they age.

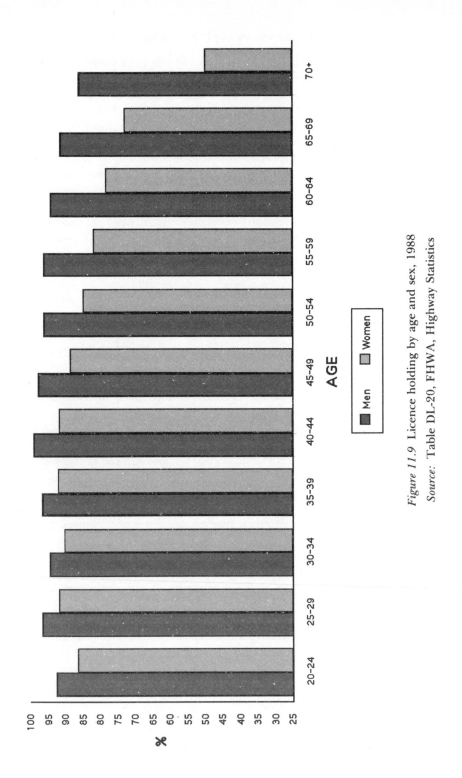

Figure 11.9 Licence holding by age and sex, 1988

Source: Table DL-20, FHWA, Highway Statistics

While such generational effects are precisely the issue raised here, these data only show us differences, not movement, between having and not having a driver's licence and between having or not having a spouse.

Stranded widows and other women without licences

While women who are over sixty-five today are far more likely to be non-drivers than men in the same age group, the percentage of older women with licences has increased considerably since the 1950s. Figure 11.9 shows how great is the recent licence-holding difference between older men and women; only slightly over 70 per cent of women over seventy had licences in 1988, compared to almost 95 per cent of men over seventy. However, Figure 11.9 also makes clear the very high licensing rates among younger women; the 90 per cent of the cohort of women who were fifty to fifty-nine and drove in 1988 will all be elderly before the end of this century, substantially reducing the percentage of un-licensed elderly women.

Although all elderly women without licences may suffer restricted mobility, there has often been a special concern with the married non-driver, who may face a sudden and drastic loss of mobility at her spouse's death. Since elderly women without licences make a high percentage of their trips in private vehicles,[8] it appears that they rely on driving spouses and family members for a substantial part of their travel (Rosenbloom 1988a).

Table 11.2 shows the vast difference in actual trip-making between women in couples and those living alone. Although the data on single

Table 11.2 1983 per capita vehicle trips annually by retired people living with someone, and living alone

Trip purpose		Retired couple		Single† retired adult	% Fewer trips
	Total	By men*	By women*		
Business	69	40.8	28.2	0	–
Shopping	2,124	1,257.4	866.6	275	68.3
Personal	1,024	606.2	417.8	137	67.2
School/ Church	357	211.3	145.7	51	65.0
Medical	250	148.0	102.0	28	72.5
Visit	659	390.1	268.9	142	47.2
Pleasure	46	27.2	18.8	17	9.6
Other social	1,104	653.6	450.4	148	67.1
Other	152	89.9	62.1	5	91.9

* Estimated
† Roughly 9% may be men

Source: Calculated from unpublished 1983 National Personal Transportation Study Data

retired adults in the table include some men, it is clear that women who live alone make substantially fewer trips for all purposes than those who live with their husbands. In most major trip categories, single women make at least two-thirds fewer trips than comparable married women. The large gap between those with and without access to a household driver may be far worse than Table 11.2 initially suggests because, of course, some percentage of the single retired women also drive, 'inflating' the per capita average.

However, controlling for a driver's licence, the gap between those with and without spouses narrows considerably. Table 11.3 compares the tripmaking behaviour of women of various ages with and without a driver's licence, and with and without a spouse or male household driver. The table clearly shows that it is the holding of a licence, not the presence of a male driver, that makes the most impact on women's travel behaviour. For women over sixty-five, the gap between those with and without licences is staggering. Conversely, while living with a male driver does increase tripmaking slightly, particularly for the oldest women making the fewest trips, in no case does having a male driver cover even 10 per cent of the gap between women of comparable ages with and without driver's licences.

This analysis shows that the stranded widow is simply a sub-set of a far larger group of women who may suffer mobility deficiencies because they have never driven, or increasingly more likely, are no longer able to drive. While these data are not perfect, they show that the stranded widow, as a specific group, is less of a problem than intuitively thought – since having a male driver around is a poor travel substitute for having the ability to drive.

Given the high licensing rates among younger cohorts of women, by 2020 almost all women over sixty-five will have been licensed to drive; the stranded widow, to the extent she exists, may be a phenomenon of just one ageing generation. The most pressing problem will not be the woman who has never driven, as important as that issue is today, but the woman who has driven all her life but can no longer do so.

The analyses below suggest that such women may suffer even greater travel deprivation, not simply in a relative sense, but in an absolute sense, because they have based their housing and lifestyle choices on the access offered by the car. Without the ability to drive (or to maintain a car) they can no longer get to many of their accustomed destinations without help.

Ageing women drivers

The new cohorts of elderly women drivers are initially less likely to suffer the mobility deficits faced by those currently without licences, but

Table 11.3 1983 per capita trips by vehicle, urban and rural areas

Age group	All females with a licence	All females without a licence*	% of additional trips by those with a licence	Females without a licence who live with male drivers	% of additional trips by those who live with a male driver
31–45	2,483	1,032	140.6	1,198	16.1
46–60	2,029	763	165.9	780	2.2
61–65	1,688	697	142.2	842	20.8
66–70	1,302	482	170.1	563	16.8
71–75	1,729	927	86.5	895	−3.5
76–80	1,534	416	268.8	630	51.4
81–85	762	297	156.6	331	11.4
85+	541	105	415.2	242	130.5

* This column includes both those unlicensed women who do, and those who do not live with a male driver.

Source: Calculated from unpublished 1983 National Personal Transportation Study Data

they will eventually lose their driving ability. Moreover, these women will sharply reduce their driving behaviour as they age, partly in response to a lowered demand for travel (Rosenbloom 1988a) but partly in response to their perceived loss of driving skill (Brainn 1980). Thus, while there is a tendency to assume that driving is an all-or-nothing phenomenon, like licence-holding, there is substantial evidence that elderly drivers suffer significant and perhaps hidden mobility losses long before they give up their licences or driving. Since elderly women are many times more likely to be living alone than comparable men, it is likely that they will have fewer alternatives when they can no longer drive, all the time or occasionally.

One way to gauge both the impact of ageing on driving behaviour, and to assess whether women suffer greater mobility losses as they age, is to compare the behaviour of men and women drivers. Women drive significantly less than comparable men; in all age categories over sixty-five elderly women drive less than one-third of the miles driven by similarly aged men (US Department of Transportation 1987). Unfortunately, this datum is not very useful because all women, across the existing generations, drive significantly less than men and we cannot be sure if this difference represents lower mobility or simply a lower desire for travel, either for those younger or older than sixty-five.

Figure 11.10 shows graphically the travel differences between men and women;[9] at no age do women drive more than 40 per cent of the miles driven by comparable male drivers. One possible explanation is that women are, in fact, travelling as far as comparable men but that they often do not drive the car if a male driver is present. However Figure 11.11, which shows vehicle miles travelled (as opposed to driven) for different trip purposes for those over sixty-five, reveals that a sizeable gap exists in actual distance covered for elderly people. The figure shows that women over sixty-five travel less than comparable men as drivers or passengers: between 40 and 50 per cent fewer miles for all trip purposes, except visiting and social, than comparable men.

Table 11.4 suggests that distance travelled is not equivalent to trips made; the table shows that while younger women routinely make shorter trips, they make just as many trips as comparable men. The top of the table shows the pattern seen in Figure 11.10; at all ages women travel fewer passenger miles – using all modes including walking – than comparable men, although the disparity increases with age. However, for all ages, except those over sixty-five, women make just as many, or more, daily trips as comparable men. Conversely, among those over sixty-five, elderly men make almost 47 per cent more daily trips than elderly women.

While it is difficult to say exactly why older women seem to make fewer trips than comparable men, the pattern is most likely the result of

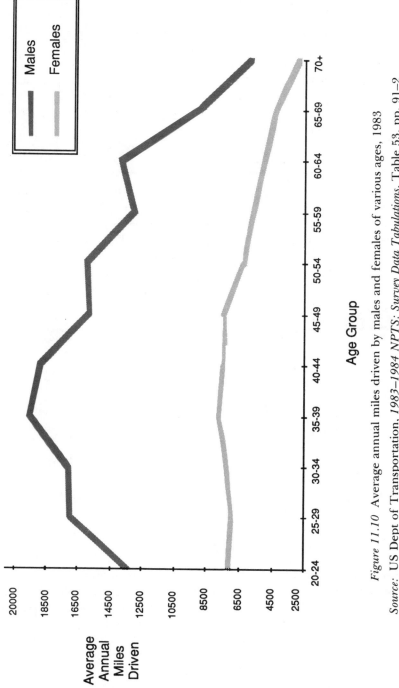

Figure 11.10 Average annual miles driven by males and females of various ages, 1983

Source: US Dept of Transportation, *1983–1984 NPTS: Survey Data Tabulations,* Table 53, pp. 91–2

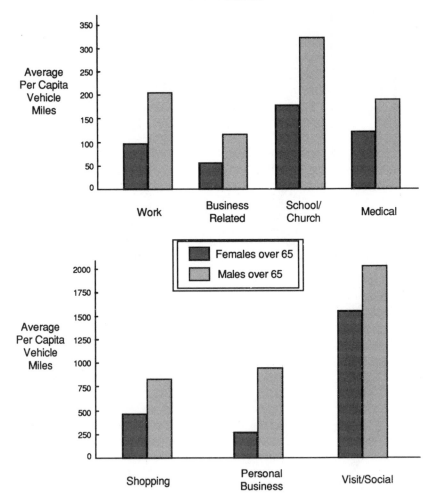

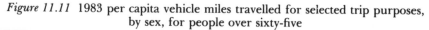

Figure 11.11 1983 per capita vehicle miles travelled for selected trip purposes, by sex, for people over sixty-five

Source: Calculated from unpublished 1983 National Personal Transportation Study Data

two factors. First, we may be seeing a simple quantitative phenomenon; there are significantly more very old women among the elderly than old men and the data groupings here are not weighted. Since trip-making naturally drops with age, the much older women in the data set reduce the average number of trips for all women over sixty-five.

However, another factor appears important – reduction in driving by elderly women still licensed to drive. There are no definitive explanations in these data, but two suggestions bear discussion: elderly women

Table 11.4 Male and female travel behaviour by age, 1983

| | 16–20 | | 21–35 | | 36–65 | | 65+ | |
	Male	Female	Male	Female	Male	Female	Male	Female
Daily passenger miles travelled/ person	22.9	21.4	32.8	29.5	33.6	25.1	14.8	10.2
% by car	80.3	82.2	89.9	86.1	74.7	84.5	90.5	88.2
Daily person trips	3.2	3.4	3.4	3.5	2.9	2.9	2.2	1.5
% by car	75.0	73.5	88.2	82.9	89.7	89.7	86.4	80.0

Source: Calculated from Tables 28 and 29, US Dept of Transportation, *1983–1984 NPTS: Survey Data Tabulations*, pp. 49 and 51.

drivers are more likely to restrict their own driving as they age than men, and, when they are unable to drive, elderly women, who often live alone or not in their own household, have no effective substitute for the trips they must make.

Studies show that elderly drivers, but particularly women, avoid high-risk situations, giving up almost all night-time or peak period driving by seventy.[10] Obviously this is at least inconvenient, but more significantly, probably leads to some decrease in trip-making, since all trips cannot be shifted from night to day and from peak to off-peak. Two Dutch researchers have suggested that the reduction in driving by the elderly of both sexes is cyclical:

> Less frequent road use leads to a loss of functions, thus leading to an extra loss of functions and of routine. The feeling of the elderly that they are no longer able to function in traffic which is tailored to the 'average' road user, and fear of their own vulnerability, have the effect that old people become even less frequent road users. A vicious circle supervenes.

> (Wouters and Wellmann 1988)

Thus it is logical to assume that some of the reduction in trip-making by women over sixty-five is a result of self-limitations, limitations which may or may not be rational. Because policymakers tend to view licensing and driving as synonymous, they are not always aware of the mobility losses suffered by those who still drive occasionally (Rosenbloom 1988b).

Elderly women drivers who lose their driving skills may be even more disadvantaged than those who have never driven, although both clearly face mobility deficits or losses. Women who have driven, particularly the 'newer' generations of the elderly, may have made a number of lifestyle decisions based on the convenience and flexibility of the private car. Increasingly they have opted for suburban living, where densities are low and distances are often too great to be covered by walking; in 1980 more of the elderly US population lived in the suburbs than in the central city (Rosenbloom 1988b) and this trend is increasing (Kasarda 1988). Suburban and rural elderly women, particularly when living alone, may have few alternatives to the private car for their daily needs.

Table 11.5 directly addresses this question by evaluating the impact of an almost ideal public transport service – an accessible vehicle with low fares, stopping within two blocks of all homes and within two blocks of all possible destinations and requiring no transfer. The table shows the average trip length, unfortunately for men and women combined, driven by those over sixty-five for various trip purposes; all but vacation trips are under ten miles and most of the rest are under five miles. Yet none of these trips could be made by walking, in under an hour, although all would consume little time in a car. Just as significantly, few of these one-way trips could be made in under half an hour by a ubiquitous and relatively perfect public transport system.

Table 11.5 Time consumed by alternative modes for average one-way trip now taken as vehicle driver for people over sixty-five

Trip purpose	Average vehicle trip length (miles)	Driving	Walking (time in minutes)	Ideal public transport
Work	5.8	10	116	33
Work-related	9.2	16	184	47
Shopping	4.4	8	88	28
Family/Personal	4.3	7	86	27
Doctor/Dentist	7.9	14	158	42
School/Church	4.5	8	90	28
Vacation	28.1	48	562	122
Visit friends	9.1	16	182	46
Social/Recreational	8.3	14	166	43
Other	5.4	9	108	32

Sources: 1983 *Personal Travel in the US*, Vol. 11 pp. E-17–E-19
Travel times were computed from 1983 NPTS, *Personal Travel in the US*, Vol. 11 p. E-18 assuming auto speed = 35 mph, walk speed = 3 mph, and average transit speed = 15 mph plus 10 minutes' walk and wait time per trip.

Even the types of trips for which these hypothesized transport options seem practical are less so on examination; it is difficult to go grocery shopping on a bus, carrying bags, for example, even if the time demands are not excessive.[11] Moreover, in reality, few elderly have access to the optimal transport services evaluated in Table 11.5; unpublished studies of Austin, Tucson, Jacksonville (Florida), and Fresno, found that between 40 and 60 per cent of the elderly simply have no public transport services at all (no matter how infrequent, or how many transfers required) within four blocks of their homes.

Other existing options are too limited in scope and operation to offer elderly women much mobility. Special public transport services, which provide much of the convenience of the private car with door-to-door or even taxi service for the elderly, are not an effective solution as currently delivered. In most cities in the United States these systems will not serve able-bodied elderly simply because they are poor, or can no longer drive, or because they have no other public transport available; only the physically disabled are eligible for service. Not surprisingly, less than 17 per cent of elderly women have ever used such services, regardless of their disability or household status (National Center for Health Statistics 1986). Recent federal legislation all but ensures that only the very handicapped will be eligible for such services in the future, so that even fewer elderly women may have such access by the turn of the century.

An important unanswered question is how great the mobility loss facing elderly women losing driving skills really is, since not all differences between those still driving and those unable to continue are caused simply by the loss of the car. People may cease to own a car or may give up a licence because illness or financial problems have reduced their desire, rather than their ability, to travel. But it is clear that some of the difference in travel behaviour between those with and without licences, and between comparable men and women, represents desired trips not made.

CONCLUSIONS

Summary

The data and analyses of the first section of this chapter show that married women in the United States have very different travel patterns than comparable men, and that single parents have different patterns than their married counterparts. Women appear to make transportation and other decisions in order to juggle successfully a number of employment, childcare, and household responsibilities; even in the Dutch context, where non-car options are safe and available, working mothers still play a significant role in their children's travel patterns (a role unmatched by their husbands).

Clearly, both single and married salaried women have transportation problems that can be differentiated from the problems faced by men and those who are not parents; embedded in all women's travel patterns are the transportation problems (including safety and security) facing their children. It seems reasonable to assume that if women continue to bear a disproportionate share of chauffeuring and other responsibilities, these travel differences will not disappear, without remarkable changes in other societal systems.

Single women are a very vulnerable group, even if they have moderate incomes; they are likely always to respond differently to the demands of their children and households. They face disproportionate pressure to alter their activities and travel patterns, not only because they are often economically disadvantaged (although the single women in this sample were not), but because they have less assistance in balancing employment with household and childcare responsibilities. This situation may leave the children of single mothers at greater risk as they try to travel independently or fail to avail themselves of societal and cultural resources in ever-dispersing communities.

Elderly women face significant problems as they age; while the majority of those over sixty-five in the future will be car drivers, initially enjoying greater mobility than comparable women today, at some point they will start to reduce their driving, lowering their mobility. Elderly women living alone are especially vulnerable, since they will have few alternatives to driving.

Policy implications

Questioning traditional responses

These findings have three specific sets of policy implications. First, the data call into doubt the efficacy of several traditional responses to the transportation problems of working mothers. Many public policies focus solely on raising income levels directly (for example, income transfers) or indirectly (for example, education, training) among working women, particularly those who are single parents. But, given the huge burdens facing these women and their children, income enhancement programmes alone cannot address the incredible time constraints or lack of societal support that create the transportation problems of either single or married female parents.

Other traditional responses call for providing public transport subsidies or traditional services to working women or elderly women; yet over 60 per cent of all women work or live in the suburbs (or in rural areas) where public transport services are poor to non-existent – and for a reason. In suburban environments, public transport is a very poor substitute for a car; data from the 1983 Housing Survey show that, on average, buses, trams, and underground services in the United States average less than 13.2 miles per hour, around one-third as fast as a car. The average suburb-to-suburb commute was 8.2 miles in 1980, so that a direct public transport trip, on an almost perfect system, would take approximately forty minutes by bus but only sixteen minutes by car

(Rosenbloom with Lerner 1989). If the real difficulties of public transport were added (unsynchronized transfers, lengthy walks at either end, etc.) transport time could easily increase by between 50 per cent and 100 per cent. Few working women can afford the time public transport takes, no matter how inexpensive; few elderly women would be able to tolerate lengthy rides for medical or other needed services.

Traditional public transport is the option of some working women, and some female elders, but fewer than 5 per cent of all women use it and the percentages have gone down steadily for both young and old women for over three decades, generally for limitations inherent in the mode. Where that is true, neither additional service nor fare subsidies will have much impact.

Recognizing the implications of emerging policies

While current programmes may fail to respond adequately to their needs, working women may suffer as a consequence of an emerging trend in transportation planning in the United States: the use of mandatory transportation demand (TDM) programmes. Such programmes – which can range from mandatory changes in working hours to banning worksite parking – can disproportionately affect working mothers. Almost no one has questioned the impact of mandated changes in employment travel on the nation's working women (Giuliano forthcoming) – although the findings of this chapter would suggest that working women with children are very susceptible to such measures.

TDM programmes attempt directly or indirectly to persuade, induce, or force workers to change transportation habits and patterns which cause traffic congestion, contribute to environmental pollution, or increase consumption of non-renewable natural resources (Orski 1986; Flynn and Glazer 1989; Ferguson 1990). These dysfunctional actions include driving to work alone, travelling during peak periods, and not using available alternatives to the private car.

Prior to the mid-1980s most TDM programmes were voluntary and – except in a few unique situations – ineffective (Roybal 1991). Simply promoting alternative commuting options rarely had much impact on travel behaviour, because so many non-transportation elements influence people's travel behaviour (Beroldo 1991; Comsis Corp. 1989). Many analysts concluded that, in order to change personal behaviour to a great enough extent to make a discernible impact, a TDM programme must be mandatory or must involve significant monetary penalties for driving alone or at all (Rosenbloom and Remak 1979; Dowling, Felham and Wycko 1991; Frederick and Kenyon 1991; Orski 1990). Today a number of states and regions have begun to implement TDM programmes which have one or more compulsory elements (Deakin forthcoming; California Department of Transportation 1990; Flynn and Glazer 1989; Orski 1990). Moreover, the 1991 Highway Bill (pending in Congress at the time of writing) makes provisions for the implementation of TDM measures as a condition of some highway and transit funding.

The growth in mandatory TDM programmes makes it likely that many working women in the United States will have to respond – adding another set of problems to those they already face in the paid labour force. Yet even low-income working women who drive alone have accepted the expense of a car because their other economic concerns (the hourly cost of childcare) or non-economic needs (the actual availability of a childcare provider matched to their work schedule) are more pressing (Davidson 1991; McKeever, Quon and Valdez 1991). Given the average time differentials between the car and all other modes – and the many time constraints and parenting demands they face – most employed women cannot easily use public transport or join carpools if they do not already do so.[12] Again, public transport subsidies are hardly likely to offset additional costs imposed on these women by mandatory changes in their work trip.

Before there is widespread implementation of mandatory measures, it is crucial to examine how the various elements of TDM programmes affect different women in the labour force. While all of these programmes have meritorious goals, it may be difficult or impossible for some women to balance their domestic and employment responsibilities under such constraints, or without the addition of other societal support mechanisms.

Fashioning new solutions

The data presented in this chapter suggest that the needs of working women and their families, and of their ageing mothers, need both short-term and long-term responses. In the main, transportation options represent the short-term responses; long-term answers lie in better public and private support services for women and children, better land-use planning, and meaningful housing options for all kinds of families.

Women must chauffeur their children because so few activities are available in their neighbourhoods or next to their schools; they must keep a car because there is no easy way to get the dry cleaning or to do the shopping at their work site or near to public transport or in their immediate neighbourhood. Mothers must drive to work so they can rush home to their latchkey kids because school hours do not match work hours, or because after-school care is unavailable or too expensive. Working women must forgo carpools because they worry that a child will become ill during the day, and few schools or childcare centres have facilities for sick children.

Elderly women face similar demands which dictate their need for a car. They live in neighbourhoods without pavements and footpaths – and nothing within walking distance, in any case. Their streets are designed for the car and pedestrians are unable to cross an intersection in one signal cycle – if they can figure out traffic signals designed for drivers. Most elderly women will age in the homes in which they were middle-aged; they have a network of doctors, friends, church, shopping, and recreation which they have built over decades – decades in which they drove easily from one part of the metropolitan area to another. Most of their normal destinations could not easily be served by a bus; who would want to travel for over forty minutes one way to see a doctor when feeling unwell? Even if elderly women chose to give up all the people and things with which they were most familiar(!), they would find few medical or commercial activities or social or human services in their immediate neighbourhoods.

Clearly transportation demands arise because of the complex inter-action of land-use patterns, housing options, and the way in which public and private services are organized and delivered. While additional transportation services and subsidies can address some of the discrepan-cies, prudence should dictate attention to the part of the transportation–land-use equation which seems to matter most – land use.

But in the short run, working women and their female elders would be best served by cars or services which act like cars – allowing them as much choice and independence as possible. Elderly women who drive should be kept in their cars as long as possible and for as long as they can demonstrate competence – through driver retraining, vehicle modifications, and financial assistance. Instead of traditional fixed-route public transport, elderly people should be provided with specially tailored fixed-route services in accessible vehicles; communities in Sweden and Great Britain have had great success with carefully routed transport services designed to provide services close to the homes and the more common destinations of elderly travellers. When they cannot use these 'service routes' and cannot or should not drive, elderly women should be offered the discounted or free service of traditional taxis; when they can no longer travel independently, elderly women (and men of course) should be supplied with specialized transport services with escorts if necessary.

Employers should be required to take a greater role in the juggling act faced by most married and single parents. Providing more financial assistance for childcare – particularly for after-school care and options for sick-child care – would probably have more impact on many working mothers than offering transportation subsidies. (On-site childcare might

241

be useful to some mothers, but employer assistance must go beyond that – mothers with school-aged children face great transportation demands as well.) Encouraging the location of banks, post offices, and, for instance, dry cleaning pick-up at work sites would reduce the need for midday car travel and facilitate the use of alternative modes.

Employers must also consider subsidizing carpools and vanpools, as well as continuing to offer subsidies to those women workers who wish to use public transport; current tax laws – which now permit employers to deduct unlimited parking subsidies but only a few dollars per month for public transport or carpool subsidies – must be changed. Employers must also be required to provide support services which will allow working women to give up their cars; guaranteed ride-home programmes (for midday emergencies) must be implemented on a meaningful scale.

Even with these employer initiatives, it is hard to see how many working women could successfully juggle a home, a job, and several children in the cities we have today without a car. It is equally difficult to contemplate the isolation facing an elderly woman who can no longer drive and so is unable to visit with life-long friends or to use doctors of thirty years' standing. To address the complicated travel patterns that working women have, we must ultimately consider the kinds of homes they can buy, the kinds of jobs they can get, and the activities in which their children can easily engage; to address both their needs and those of the elderly, we must consider how to expand the range of services and facilities offered in their neighbourhoods.

12

WOMEN, THE STATE AND THE LIFE COURSE IN URBAN AUSTRALIA

Ruth Fincher

The policies and practices of the State mediate women's experience of the life course significantly. Norms embedded in social policies make government benefits available at certain times in women's lives, usually times associated with the expected passage of their mothering or caring roles, and benefits can accrue most readily to those women fitting such life circumstances. Of course, simple (often implicit) life course classifications of women, their needs and rights, are also accompanied in state policies and practices by assumptions about women's class, race and ethnic characteristics.

Women active in collective protest encounter the State at every turn, and must often contest its norms and language. The State itself, however, is not a predictable, unchanging and consistent set of institutions and practices. In some places, the activities of agencies and employees of the State can empower women of certain life course groups at the very times that other parts of the state apparatus deny them. Particular life course groups, in turn, can be disadvantaged by government policies even as others are facilitated. It is impossible to generalize, then, about the implications of all social policies for all women across their lives in the various places they live. Indeed, some theorists have used this empirical reality as an argument against the political usefulness or validity of feminist theories of the State (Allen 1990). There remains a critical tension in feminist (as other) state theory as to whether broader theories of the gender relations of the State are bound to neglect critical issues of diversity amongst women and amongst state apparatuses and practices, or whether a broader theoretical position may yet encompass difference in the empirical insights it generates.

This chapter seeks to demonstrate life course perspectives underlying certain social policies of the Australian state, and their implications for the daily lives of women in cities. Its evidence is about women and governments in Melbourne, a metropolitan area of over four million people. The central arguments are two:

243

1 That many social policies envisage women as mothers, and 'reward' women with certain social policy 'advantages' if they are in marital relationships with, and dependent on, men. In particular, women are perceived in much social policy as at the stage of heterosexual marital relationships where they have small children; they are perceived to construct their daily lives around the demands of care for those dependent children. This life course view of women, which results in government policy 'assistance' for women in particular life circumstances, is one which might disadvantage women at other life course points, relatively speaking.

2 The implications of complying with social policy expectations of women, or not, vary by class. A woman is more susceptible to the disadvantages that can result from *not* fitting policy expectations if her resources (of both income and capacity to control her circumstances) are limited.

Analysts of a range of Australian social policies have noted particular life course views of women and the activities women are expected to engage in at those life points. Australian immigration policy, for example, particularly in its family reunion criteria, values women in their 'breeding' years (de Lepervanche 1989). But the way this policy preference has been exercised is that women in nuclear families have been favoured over women in extended or *de facto* family arrangements (Hugo, 1990: 57). Disadvantage may accrue to particular groups of women applying to migrate to Australia, because of these rules.

Australian children's services policy has also shown its preference for maternal women. Brennan and O'Donnell (1986), for example, note the tussle between kindergarten and child daycare provisions in the history of Australian children's services and the preference shown by governments, until very recently, for kindergartens. With kindergartens, mothers are generally at home and can deliver and collect their children after the short daily sessions; with long day childcare, services are timed more to coincide with the commitments of working mothers. Gender and life course preference has also been documented in Australian housing policy, in which home ownership by married heterosexual couples has long been supported in preference to other tenure forms and household types (Watson 1988). Clearly, the State interacts with people in many ways and so helps construct life course experiences: in taxation and superannuation policy, in transport, recreation, education, medicine, social services and in the support of certain sectors of production. Each sphere of government policy and practice has life course and gender implications which need to be made clear.

This chapter will consider one instance of the intersection of women with the State, and the significance of women's life courses in this intersection. The example is of women in the paid labour force who are users of local community services in urban spaces. The access they have to these services varies according to the particular urban municipalities in which they live. From this case, it should become clear that local community services, which are vital to labour force participants with dependants, are useful primarily to women of particular class, ethnic and life course groups; also that gendered spaces are being produced in cities as a result of processes like the locally varying access of working women to services. These spaces will function well as living spaces for some groups, and not at all well for others.

I will concentrate on identifying some of the characteristics of the policies and state practices rather than substantiating the claims made with opinions gathered from the women affected. In this, the chapter represents a call for further feminist research about the effects of the life course assumptions of the State.

Before considering the matters listed, I will briefly review feminist perspectives on the State to show how these suggest implications for women at different life course stages. Though feminist approaches to the analysis of the State now consider class, ethnic and racial diversity amongst women, they have paid little attention to life course differences. However, if we take up the spirit of feminist enquiry into the State, which is to encompass the diversity of women's experience, then questions about the State and women's life course are readily posed. Such questions direct the material about women in Melbourne that follows.

WOMEN AND THE STATE: IMPLICATIONS FOR THE LIFE COURSE

In the last four decades, the welfare state has expanded to regulate many aspects of social life in western societies. The bureaucracies of the modern State are major players in demarcating the life course stages now experienced so sharply. In a useful paper, Karl Mayer and Walter Muller set out to show how the individual life course and the evolution of the modern State have occurred together, and also 'that the State created important structural prerequisites for the coming into existence of individual life histories' (Mayer and Muller 1986: 224). Their concept of the modern structured life course is that individuals' lives are a set of phases that both social institutions and the individuals themselves identify and construct; these phases are usually separated by transition points which we define normatively. The instruments of state action identified by Mayer and Muller as the means through which these outcomes have been created include regulative norms and age stratification (which influence the formation of life course stages when they are combined with government services or financial contributions); entitlements and standardization of events (like social insurance, which one is entitled to receive according to being in work or not in work); and modes of bureaucratic organization (including language and the bracketing of services into professional specialisms).

This account does not draw out the gender subtext of the State's influence on the definition of life courses, nor the presence and complexity of the activism which has accompanied the large-scale transitions of society to which it refers. But there are gender relations in the State's involvement with life courses, and the ways the various arms of the State have defined their roles and positions continue to be contested. Indeed, debates about the State amongst feminists have emphasized the significance of its forms and processes for women in different circumstances, and the actions, of women in particular, with respect to these. These contributions have ranged around the following issues, which have significant implications for the stages through which women live their lives (or at least our understanding of how to interpret the significance of the State's activities for women's experience of the life course):

1 How the State 'constructs' the category and marginalized group of 'women', quite apart from the diversity within that group.
2 Whether the increased presence of the State, on balance, causes additional difficulties for women, who must add on to their common dual roles of paid work and domestic unpaid work in the home the further task of coordinating their interaction with whatever state services and provisions they use.
3 The significant types of interaction women activists have with the modern State, that seem to channel its outcomes in ways with which many are dissatisfied.

The first of these matters, how the State has 'constructed' women as a group, has been explored in the context of the characteristics of citizenship by feminist political theorists, particularly Pateman (1988). Pateman argues convincingly, from a close analysis of the social contract that underpins the capitalist democracies, that though women have been incorporated into a society in which the freedom of all 'individuals' is guaranteed, in fact this freedom (where, we might add, it occurs at all) only accrues to those persons in male bodies. The State, in this analysis, is patriarchal at its very core. Shaver (1990: 9) pursues the implications of 'the rights of legal personhood', as she terms them, for the production of gendered social policies, but goes beyond Pateman in identifying issues related to the life course. She notes the significance of the heterosexual marital couple in the financial provisions of the Australian state, dating from the early twentieth century when minimum wage levels were determined by awarding men a 'family wage' appropriate to supporting a wife and three children, but awarding women little more than half that amount (Shaver 1990: 12). A similar gender and hetero-sexist logic prevents the second earner in a marital household from 'access to income support in disablement, sickness and unemployment' (p. 14), a condition to which homosexual couples and indeed unmarried couples are not subject. Even as the State moves towards some ideas of 'gender neutrality', 'the deeper gender logic of financial partnership of the couple continues unchanged, . . . in the continued basis of income security on the heterosexual marital couple' (Shaver 1990: 18–19). As Pringle and Watson (1990: 236) point out, Shaver's work shows how the State takes from women as individuals and redistributes to them through the tax system as wives and mothers.

The life course bias of these policies consists in the way that they structure advantage for women (and men) in particular circumstances, providing incentives for them to make certain choices at points in their lives where they have distinct family commitments and social and economic needs. Pringle and Watson (1990: 236) claim that the welfare state 'has no interest in women as non-family members, as women outside their domestic and mothering role'. This needs to be closely investigated in the case of particular policies and government practices. Single women old or young, lesbian women, women without children, women in the paid labour force in all these categories, seem less directly favoured by some subsidies and services. What these groups of women have in common is their absence from a common life course 'stage' in which active mothering in particular domestic circumstances is their core activity. It can be hypothesized, then, that various policies of the State may actively encourage dependent marital status and mothering.

The second issue explored by feminist analysts of the contemporary welfare state is the oppressiveness of that State for women. Much has been written to indicate the ways that women's lives remain those of dependants, as the tasks of social reproduction in which they remain engaged are regulated and facilitated by the State. So, Balbo (1987) has likened women's lives to 'crazy quilts', describing their activities as 'making-do' within multiple schedules established by others – paid workplaces, families, government service-providers. Hernes (1987) points out that it is the incompleteness of the State's entry into reproduc-tion that causes such problems for women, who must do the bulk of the servicing work that the State fails to do.

An uneasy partnership between women and the welfare state exists, then, to accommodate some of the work of reproduction as women have entered the paid labour force in large numbers. This remains oppressive of women because it depends on the entrenchment of women's dual roles as wage earners and primary carers (Borchorst and Sim 1987). As women bear primary responsibility for the care of society's dependants, they remain dependent themselves on help in providing such care.

Others have argued against the view that developments in the welfare state are utterly oppressive. Jones (1988: 25) proposes that the 'oppressed group model' for defining women's political roles should be discarded in favour of analyses of the specific and varied ways women 'have wielded power, been in authority, practised citizenship and understood freedom'. And Piven (1985: 265), noting the 'almost categorical antipathy' of recent feminist literature to the State, points out the irony of women intellectuals scorning 'dependent' relations with the State at the same time as women activists use its resources and influence. Rather than seeing women's complex relationships with the State as a new cause of powerlessness, Piven (1985) interprets this historical shift as giving some power and protection to women, and as affirming and facilitating the nurturing they continue to want to do.

What are the life course implications of this view of the partnership, willing or otherwise, between women carers and the State? Quite apart from whether women want to nurture or not, the partnership needs to be examined as to whether it offers women at different life course stages more or less opportunity to combine caring with other activities. It is also crucial to see how this varies between women of different classes and ethnic and racial groups.

The third concern in examinations of women and the State has been the specific interactions of women with welfare state institutions, indicating how these interactions generate experiences and outcomes with which many women are dissatisfied. Many such interactions centre on the ways that bureaucracy, the mode of organization of large state agencies, alters and dilutes the demands of women activists.[1] Insights made into women/state interactions include the tensions within the women's movement between feminist bureaucrats and others, and how demands are often transformed once they have been expressed in ways consistent with government criteria and ideologies.

When activists have lobbied the State for gains for women, they often end up using state resources to develop the programmes requested. The responses of the State to feminist demands have often led to the 'institutionalization' of women's issues (Barnsley 1988: 18) and with this their transformation. In particular, the meaning and significance of feminist demands are altered as they come to be expressed in terms that comply with the existing regulations and expectations of government funding agencies (Barnsley 1988; Monkman 1988). An analyst of the Canadian childcare movement, for example, has noted that with the institutionalization of childcare within the Canadian state, 'the critical perspective which demanded transformed relations through the social-ization of childcare in the 1970s is now barely a whisper' (Prentice 1988:

59). The degree to which the State can co-opt demands made of it also depends on the efficacy of those making demands in actually entering the discourse of the State, and using it for their own ends. Real access to the State's discourse is a matter of class capacity – it is likely to be achieved by middle- and upper-class individuals and groups, and not by others.

Considerable tension is reported in the Australian literature between different groups of women interacting with the State. Central to the development of the women's movement in Australia in the last decade have been 'femocrats', women who are senior government bureaucrats and represent the interests of women in high-level decision-making. These are middle-class Anglo women, for the most part. They are generally not older than middle-age, for they have risen to their positions since the election of the social democratic Labor (federal) government in 1983 (see Watson 1990; Yeatman 1990; Franzway, Court and Connell, 1989). They consider it possible to achieve major reform in the interests of women from within the State: 'femocracy' is a political strategy. And yet femocrats are criticized and resented by others in the women's movement – because they have highly paid jobs which keep them remote from women in more marginal economic circumstances, because their allegiances must be to their public service employer as well as to women and because as middle-class, Anglo professionals they cannot truly represent other groups of women, including, for example, those of non-English speaking backgrounds. They, in turn, resent being positioned in this way.

If women of a certain class and ethnicity monopolize decision-making about women inside the State, potentially their presence will enhance the class capacity and power of women outside the State who share their interests and characteristics. These might include the life course related concerns of middle-class women – in particular, those juggling to combine care of small children with professional, time-consuming jobs. The combined effect of women within the State drafting policies that fit their own experiences, and their understandings of women's daily lives based upon what is expressed by activists from outside the State who share their concerns, might be the development of policies most benefi-cial to middle-class, Anglo, professional women. The choice of some women to advocate women's interests within the Australian state, then, has generated tensions within the broader women's movement. Some analysts regard femocrats as having made considerable gains for all women: funding for community services, shelters for women, equal employment opportunity policies and so on. Others are more cynical, and critical. Unsurprisingly, a homogeneous 'women's interest' has not emerged (nor can it, according to Jonasdottir (1988)).

If the discussions of women and the State were extended beyond their current foci still mostly on women as one group, despite some acknow-ledgement of diversity, to consider life course questions, the following issues might emerge to guide further enquiry in particular policy and activity contexts. Do state policies and practices privilege certain groups of women over others? Do certain groups of women engage more in activism, or interact more successfully with the bureaucracy? How do state policies interpret or develop ideology about women's caring that might affect women's actions at times of their lives when they have particular responsibilities? Crossing any systematic analysis of the bene-fits and constraints associated with different life course circumstances, however, should be consideration of how class capacity affects women's vulnerability to the State's dispersal of costs and benefits. So, too, there needs to be appreciation of how some life course-dependent interactions with the State are spatially explained – the benefits of some govern-ment provisions are available to people in certain localities and not in others.

WORKING WOMEN IN DIFFERENT LIFE COURSE STAGES

Women who work in the paid labour force are often in partnership with the State to enable them to care for their dependants whilst they are employed outside the home. Frequently they use locally provided, government-subsidized, community services – the precise forms of services and eligibility of women varying between different places, of course (e.g. Sassoon 1987; Fincher 1989, 1991).

How is the partnership between women worker-carers and the State constructed in Melbourne's municipalities? First, it involves certain parts of the state apparatus having responsibility for particular government programmes: for example, the federal government subsidizes child daycare places in government-funded centres, targeting expansion of such places to designated 'needy' areas. State and local governments can provide buildings for childcare centres, whilst the federal government's subsidies and other grants made directly to centres provide recurrent operating funds. Local governments run federally-subsidized family daycare schemes, in which carers take several children into their homes to care for them there. Non-institutional care of elderly dependants is a 'community responsibility'; local government delivers federally-funded services (for example, assistance with laundry and house clean-ing) to the aged in need in their homes, and limited day centres exist for the aged. State, federal and local governments contribute to the range of services offered. Women, as 'partners' to the State, use these community services in lieu of caring for dependants in the home full-time. Not all local municipalities provide the same services to their

250

residents, however, in form or extent (see Fincher 1991), and these government fiscal arrangements are not static.

Second, the joining of local women to the State (or, according to your perspective, their dependence on it) by these financial provisions rests on their activism and vigilance within localities and in forums that affect local governments, to ensure that services are made available as far as possible in the forms desired. The terms in which governments interpret the community service 'needs' of women are thus continually contested, resulting in considerable spatial differentiation in policy and implementation outcomes, as well as in the characteristics of the struggle itself.

Within Melbourne, municipal variation in the community support of workers with dependants includes:

1 The extent of services – not only the number of people served but the quality and frequency of services offered.
2 The supply policy – for example, whether the local services are provided by local government employees or volunteers; whether local councils establish child and aged daycare centres or whether these are private, for-profit establishments; whether local councils manage such centres or turn them over to volunteer committees of management and fund-raisers.
3 The local arenas of conflict within which the extent of services and the local supply policy are debated. These vary with the amount of consensus between local government community service bureaucrats and the elected local councillors, and with the types of interaction between community groups, councillors and bureaucrats (for example ongoing consultation versus local government response only when there are eruptions of local protest).

Detailed examples of these variations are given in Fincher (1991). What they make clear is that local spaces of daily interaction with the services of the State are formed through people's contest with the State, as well as their experiences of access to its precise provisions. Furthermore, these are gendered spaces because the access of women to community services either hinders or facilitates the capacity of those women to undertake paid employment outside the home, or indeed to embark upon a range of other activities. The class capacity of women will determine, to a considerable degree, their vulnerability to local levels of service provision. If they have the resources to pay for privately organized care in lieu of using government services, or if they have the time, connections and capability to lobby local governments effectively to improve local government services, then they will avoid the worst effects of inaccessible or inadequate services.

These gendered spaces are not rigidly bounded and homogeneous physical blocks of land. Rather they are assemblages of social practices that are at once workplaces for women, homes that are variously 'serviced' by childcare and the like, and sites of political activism that influence the levels of control people have over their daily life circumstances. Of course, what are 'serviced homes' for some women are the workplaces of other women. The nasty visions of an adequate childcare system for young middle-class women relying upon underpaid female childcare workers (often of the same age but unable to find alternative work outside their home), or of the extension of domiciliary care for the dependent aged relatives of women who wish to retain paid employment, depending upon volunteer or poorly paid local home help workers, are disquieting.

Urban planning often reinforces stereotypes about the life course stages of women in particular types of location, even if these stereotypes are at odds with the characteristics of actual residents. Such planning also helps to cement and delimit those stages in the daily lives of residents. Land-use regulation of suburbs in North American metropolitan areas is perhaps the paradigm example of this: local zoning laws last upheld in the US Supreme Court in 1974 have made possible the interpretation that single family homes are to be occupied by just that, single nuclear families. As Perin (1988: 40) notes, a family of this sort depends on the presence of husbands and fathers for legitimacy – widows, divorcees and spinsters often experience stigma and hostility, because they are aberrations in this suburban setting. The 'perfect' suburban household comprises a married couple with dependent children – life course characteristics are implied in the circumstances of the people who are residents of suburbs, as are gender relations.

Though services to support the dependants of labour force participants vary locally, if one is talking about the life course implications of service provision for working women it is possible to generalize about the level of government-funded community support services available to women in paid work in Melbourne, who require care for dependants. Information on the actual users of such services (though limited) permits further understanding of the variations in access to these services by class. The general situation is that the parent of a small, able-bodied child has some possibility of remaining a labour force participant. Some such women will gain places in government childcare programmes for their dependants, others will make arrangements with family daycare workers. If you have a dependant who is not a small able-bodied child, the options are much fewer. The particular life course implication of this is that women in their 'middle years', when they have aged dependants (often as well as dependent older children) are unlikely to be able to retain regular paid employment if they must rely on the State to care for their dependants.

252

The women in Melbourne best served by those community services that support the dependants of labour force participants are of certain class and ethnic groups, as well as of certain life course stages. Least vulnerable to variations in services provision are those with the resources to be impervious to any shortfall in government provisions. These women, with higher-class capacity, are excluded from the picture painted here. But within the group who could be helped by such services and could not replace them by private arrangements if necessary, certain groups are favoured. Middle-class, Anglo women at the stage of their lives when they combine paid work with the nurturing of small, able-bodied children are favoured by existing state policies and practices.

Data on the users of government childcare services illustrate this. (Day centres for elderly dependants and municipal home help schemes do not keep records of the occupational status of their clients' primary carers, which might have been compared with information on childcare users. This situation reinforces my point that the capacity of carers of the aged or disabled to work in the paid labour force is an issue simply not addressed in policy circles.) The users (both men and women) of government childcare centres in seven Melbourne municipalities whose records were examined in 1988 have the following characteristics. The municipalities are inner, middle and outer suburbs of the metropolitan area. In all but the outer suburbs the proportion of users in the professional and managerial occupational groups is higher than the proportion of these occupational groups in the municipal population as a whole, whereas the occupational group of labourers and machine operators is lower. Though they are far fewer in number than two-parent household users, single-parent families are over-represented as users of childcare centres and family daycare schemes everywhere in the municipalities studied. The single-parent users generally work full-time. In higher-income areas, however, the two-parent households using childcare services generally comprise one full-time and one part-time worker (Fincher 1991).

What are the reasons that only certain parents of young able-bodied children are likely to make use of local childcare services? The services are costly, with fee relief subsidies available only to those on very low incomes. (It is worth noting that the incomes of some women are so low that they are eligible to receive fee relief even whilst working full-time in paid employment.) Services in centres are also restricted in their hours of opening (for example, 7.30 a.m. to 6.00 p.m. Monday to Friday). Because most centres prefer to fill as many places as possible with full-timers, access for part-time users of childcare centres may be restricted.

Within the groups of potential users of these children's services, one group, non-English-speaking women, who are usually immigrants, is particularly disadvantaged. Often having few material and cultural resources, coupled with communication difficulties with other members of lobbying groups as well as with government bureaucrats – and sometimes with cultural predilections that may be at odds with the use of government childcare centres – they exhibit reduced class capacity to gain access to this form of state support for dependants. I include non-English-speaking women as a group with lessened class capacity, rather than as a group whose primary characteristic is culturally different preferences. A recent survey in Victoria found that only 7 per cent of the childcare places it identified involved 'parents with a low degree of English language competency'; it also noted that an earlier national survey found that 83 per cent of family units in which at least one parent was born in a non-English-speaking country used informal rather than formal childcare arrangements (Victorian Department of Labour 1989: 75; cf. Rose this volume). The reasons for this pattern are disputed. Women of non-English-speaking background often find the format of children's services inaccessible and culturally inappropriate, and the cost high, especially if they are newly arrived migrants struggling to establish themselves. Activists amongst ethnic minorities claim such women have a different perception of what constitutes appropriate childcare, and are unwilling to allow their children to spend long periods with strangers:

> This is an acknowledged reality. But because the 'specialness' requires a different approach, untried and unknown to many of the predominantly Anglo-Australian bureaucrats and service providers, child care services for families from NESB [non-English-speaking background] are often reduced to the provision of workers from NESB, who introduce the existing concepts and services to the families, in their own language, of course.
>
> (D'Mello 1990: 3)

These activists claim that Australia's so-called multicultural society should provide services that are *really* different for the various ethnic groups, rather than services which share or have acquired Anglo assumptions dressed in non-English words. Others claim that perhaps a more important factor underlying the low use of childcare services by people of non-English-speaking background is cost. Since immigrant women are highly likely to be in paid employment and in households in which another adult is also receiving a salary, it is unlikely they would be eligible for fee relief. Childcare fees of around A$110 and more per child per week are common. It is little wonder that new migrants commonly take on split shifts, to avoid having to purchase childcare as they establish themselves in their new country (Fincher, Webber and Campbell 1989). (Of course, split shifts may also be the only resort of those who must undertake shift work at hours when childcare is unavailable.) Further, places are not available for all children requiring them -- an Australian Bureau of Statistics survey in 1987 established that 499,000 families with all parents in paid work used childcare (formal, informal or both), but other analysts have estimated an unmet demand in the order of 125,000 places for all Australia and 6,050 places just for the state of Victoria (Maas 1990: 60).

In Melbourne, childcare places are certainly more numerous in the inner city than the outer municipalities (Burbidge, 1990: 30). Indeed, the lack of such services is now widely known: in the biggest metropolitan areas of Australia, suburbs on the urban fringe have become almost synonymous in the public mind with newly married young couples with small children, immobile and struggling to become home owners (see Foy and Crafter 1989). As Richards (1990: 255) says in her study of such a context in Melbourne: 'The image of suburban loneliness portrays young women trapped at home, not alone but burdened with young children; the suburbs as rows of little boxes whose contents are families.' Research has documented the difficulties of women in these circumstances, particularly noting the impacts of lack of social services and adequate public transport (see Coombridge *et al.* 1990; and Rosenbloom this volume) at a time of financial stress for many households.

Many suggested policies for improving the circumstances of urban fringe residents, particularly women, centre on extending childcare or children's services and public transport so that women can better juggle journeys to work, shopping and care for children. Influential amongst these at present are the statements of the Prime Minister's Social Justice Taskforce about 'locational disadvantage', the latter defined as disadvantage occurring for people because they live in areas deficient in physical and social infrastructure and access to jobs, training and recreation (Australian Government 1990–91: 9). Though these statements are broad, the major examples always concern urban fringe municipalities, and always young people requiring childcare. For example:

> policies . . . will be less effective if a person's housing location and the associated transport links, limit access to childcare, education and training facilities and jobs [and] poor public transport is a particular problem for low income people without cars and can be especially difficult for women in these situations if access to childcare and work is required as part of the same journey.
>
> (Australian Government 1990: 1, 2)

Much policy discussion of the outer suburbs currently seems to consider them to be home primarily to couples of low class capacity, young and with children. Planning to rectify the problems incurred by these people focuses on their life course requirements, childcare in particular.

While those services are desperately needed by such people, the focus suggests, first, that caring women, when they are combining caring with paid employment (or trying to), are generally envisaged as young mothers with small children, and are supported in that role; and second, that older women, caring for different groups of dependants, are less readily perceived to occupy the outer suburbs. In fact, increasing numbers of older women, and women in their 'middle years', live in the outer suburbs (Burke and Hayward 1990). The difficulties of carers of older dependants need to be addressed on the metropolitan outskirts in particular, as well as in other locations, as the following figures attest.

The labour force participation rates of married women in Australia between 1984 and 1989 show the largest increases in the age groups 35–44 and 45–54. These age groups are expected to exhibit the largest increase numerically in labour force participation into the mid-1990s (Maas 1990). It has also been pointed out that the participation rates of married women aged 55–9 and 60–4 years, the prospective 'informal' carers of their daughters' children, will increase by four and three percentage points respectively, to 36 and 17 per cent, thereby further reducing the pool of family carers available to working mothers (Maas 1990: 62–3; cf. Rose this volume).

This situation may have implications for the scarcity of childcare places. It also has implications for the caring responsibilities of those women themselves – women older than forty are those who will experience the demands of care for older dependants, often parents or spouses. If their labour force participation is increasing, what of their caring responsibilities, and what attention is being paid to their needs in this context? These questions are the more significant because the Australian population is ageing; insofar as this represents prolonged life for aged people who incur some disabilities, the tasks of caring for the dependent aged will increase.

The answer is that not much attention is presently being paid to women in their 'middle years' as labour force participants who require assistance with their caring responsibilities. As one research consultant recently stated (in personal communication with me), everyone assumes that such carers will give up paid work. McCallum and Gelfand (1990) found, in their discussions with women from different ethnic groups in Australia, only a very few individuals able to combine caring with some paid work outside the home, and quite a number who had indeed stopped work in order to care for ageing parents. The emphasis of government, non-institutional programmes of assistance to the aged has been on 'community care', which in Australia means care by women in the home (McCallum and Gelfand 1990: 1). This echoes the situation elsewhere – as Hess (1985: 328) remarks of the United States, homecare of the elderly generally means old women receive care and younger women give it.

Table 12.1 Aged care services provided by or affiliated with selected Melbourne local councils, 1988

Service	Inner municipalities		Middle municipalities		Outer municipalities		
	Essendon	Camberwell	Footscray	Sunshine	Werribee	Knox	Lillydale
Meals on wheels. No. meals delivered (per day)	300	963	200	260	63	246	115
No. staff							
– paid	4	4	13	12	5	7	2
– volunteer (rostered/day)	6	62	0	0	3	n/a	7
Home help. No. recipients (once a week)	600	600	530	600	300	700	277
No. recipients (once fortnightly)	300	650	60	150	0	0	197
No. workers (f/t, p/t, casual)	62	110	55	87	15	68	60
Special home help No. recipients	50	80	5	37	11	140	45
No. workers (f/t, p/t, casual)	12	13	5	n/a	incl. in home help	18	17
Day centres (No. council-sponsored)	0	0	0	1	0	1	0

Source: Fincher, 1991.

Such care can be either in the carer's home (often the home of the dependent person's daughter) or in the home of the aged person him/herself. Carers can also be domiciliary (or 'home help') workers supplied by local councils, who are paid employees or volunteers (see Fincher 1991), but it is almost always the case that a primary carer, usually a family member, carries out most of the caring work. The extent of locally provided services makes this clear; see Table 12.1, which describes the extent of services in seven Melbourne municipalities. In 1986 at the last Australian census, the number of residents over sixty years of age in these municipalities was greatest in Camberwell (almost 20,000), followed by Essendon and Sunshine (between 10,000 and 13,000), then Knox, Footscray and Lilydale (5,000 to 10,000 each) and finally Werribee (fewer than 4,000).

Domiciliary care, directed mainly to frail and single elderly people and couples living alone, is the main form of support for the dependent aged. Whilst meals might be delivered once a day, visits for other purposes by domiciliary workers to the households of the dependent aged are rarely more than once-weekly (see Table 12.1). Some municipalities stipulate that aged dependants may not receive such home help if their primary carer is in paid employment. Some municipalities have day centres for aged people, but attendance at these is rarely as much as two sessions per week.

Since the mid-1980s, the federal government has paid a carer's pension to those who meet certain stringent conditions: the person cared for must be a severely handicapped aged or invalid pensioner; the carer is subject to an assets test; care provided must be constant. There is no time limit on receipt of the carer's pension, and its amount and general conditions are the same as those for the age pension – both are non-contributory, means and asset tested, and are poverty alleviation devices which ensure a very modest income for the groups to which they are targeted (see Department of Social Security *Annual Report* 1989–90; Rossiter 1985). Again, this does not have carers working in the paid labour force in mind at all. One draws a similar point from the projected increases in federal assistance for carers over the next four years, which are to 'ease the strain upon the increasing numbers of frail and elderly carers who support older people with disabilities in their family homes' (Australian Government 1990–91: 7–8).

Some innovative programmes in respite care have begun; in particular, the Multicultural Respite Care Project in Adelaide has been evaluated as a success from the users' perspective (Barnett 1988). But this programme was small, with 115 carers being relieved by bilingual paramedical aides who spent time with the dependent person. The costs to families and carers of the shift to community-based care of the aged has not received much official recognition (Barnett 1988). The federal government's Home and Community Care programme, which funds a large portion of the costs of local aged care schemes, now has the care of carers as one priority. Introduction of wide-ranging support programmes seems unlikely in the near future, however, particularly since the allocation of funds from aged care resources to *carers* of the aged, rather than directly to the aged themselves, is a difficult political point to negotiate – both within government and within families. Current policy discussions further suggest that any schemes initiated will emphasize respite care, rather than replacement of carers for sufficiently long periods to permit them to retain employment outside the home. A recent note in the journal of the Australian Institute of Family Studies, for example, presents statements by researchers about unpaid caregivers. One researcher suggested from a survey that carers' psychological well-being should be improved and that there needs to be 'development and evaluation of education and support programmes for Australian caregivers'; another claimed that 'families with severely intellectually disabled children are in need of personal support, skilled counselling, "service brokerage" and a "cash fund"' (*Family Matters* April 1990: 45). Whilst it is true that many women find fulfilment in caring for aged or disabled dependants, many also need or wish to work in the paid labour force. Respite care is vitally necessary to help ease the tensions of such caring relationships (see Marsden and Abrams 1987; McCallum and Gelfand 1990), but more sustained support is necessary to underpin women's employment possibilities. Of course, the capacity of different women to avoid reliance on the limited support services of the State varies with income – women with higher incomes have the capacity to purchase non-government care for their dependants, while those on lower incomes do not.

The partnership between the State and women in their 'middle years' who have aged dependants, then, currently gives far fewer options for labour force participation than does the partnership between the State and women with able-bodied young children. The social policy philosophies underlying the services provided are different. Childcare is provided for the benefit of working and needy parents as well as for children, whereas aged care is directed solely to the needy aged. When examined against the requirements of a carer in the paid labour force, services for the care of the dependent elderly are much more fragmented than children's services.

For the dependent aged, use of local community services is more difficult if these services are culturally inappropriate. Though the experience of caring and its precise costs vary amongst ethnic groups of non-English-speaking background (McCallum and Gelfand 1990), the picture across groups is overwhelmingly one of a major work increase for women when they become carers. The caring responsibilities of women with such dependants are often greater, because they must substitute their own caring work for government services which cannot be used even if they exist. A recent survey of the characteristics of aged migrants found that a significant proportion of those aged seventy-five years and older had arrived in the past few years under the family migration programme; it found that the English language proficiency of aged migrants was low and that many of these older migrants did not wish to move from their current homes to nursing homes or hostels, as the Australian-born and English-speaking migrants often intended (Australian Council on the Ageing 1985). Yet we know that migrants who have worked in Australia for long periods, often in manufacturing industry, have higher levels of ill health than the population as a whole. This, combined with inadequate English, can reduce their mobility and increase their dependence on others for support (Dollis 1989: 14). One of the two major current concerns of immigrant women, then, has been listed as the care of their elderly, given the scarcity of culturally appropriate aged care services like nursing homes, and the difficulties of staffing community care programmes with bilingual employees (Seitz 1989: 4–5).

Despite these facts, and the growing number of migrant aged in Australia, service provision for the aged generally assumes its users are of English descent (Pensabene 1987). Further, service provision for those of non-English-speaking background quite fails to recognize the effect of language barriers on service use by migrants, and retains the mistaken but comforting assumption that aged migrants have large extended families which 'look after their own'. Instead, the aged migrants often have fewer prospective family carers, because many of their relatives remain in their home country (Rowland 1986).

For women in their 'middle years' who happen to be immigrants of non-English-speaking background with aged dependants, caring responsibilities can be even greater than they are for Anglo women, who can make at least some use of existing community services. Their class capacity – a combination of material resources and ability to interact with the State and lobby groups effectively – can be limited, compared to the native-born.

Though the State's provisions for women caring for aged dependants vary spatially, this variation is overwhelmed by the fact that provision of services permitting such women to participate in paid work outside the home are uniformly utterly limited. From Table 12.1, for example, it is clear that even those municipalities with more extensive provisions for the care of the aged have not organized the relevant services to give the primary carers of service recipients large blocks of time each week in which they might retain a job. In these circumstances, the class capacity of women caring for aged dependants – especially in the form of resources to find and pay for private homecare for dependants – becomes crucial if they are to have the option of paid work outside the home.

As well, policy documents exhibit a lack of serious concern for the paucity of opportunities for such 'women in the middle', a short-sighted view, given the projected increases in labour force participation of women in older age brackets. Even in policy discussions of urban fringe locations, presently the focus of so much 'social justice' thinking in Australia, services for working carers are primarily envisaged as child-care services, demonstrating a life course bias even in serious considera-tion of the needs of working women. This short-sightedness has a specific spatial form, at least in the case of Melbourne. Counter to the life course image often associated with outer suburbs, data indicate that the current population growth on Melbourne's fringe is spread over the whole age spectrum. Furthermore, these outer and fringe suburbs of Melbourne will experience the ageing of their populations dispropor-tionately, if the existing residents stay there: 84 per cent of the increase in the population aged over sixty by the year 2001 is projected to occur in outer and fringe municipalities of Melbourne (Burke and Hayward 1990: 12). The increasing presence of the aged in those urban spaces identified with young couples and children may soon reveal more sharply the discrepancy between stereotypes of the occupants of places and their real characteristics. Many of the aged will be women, and many will be employed – contributing to the increases in numbers and proportions of employed older women referred to already. Social policies must be altered if these women are to be supported as working carers.

CONCLUSION

I have argued that by addressing differences between life course groups and local spaces, as well as those of class and ethnicity, it is possible to reveal the ways that community services in Melbourne create possibilities for certain women to undertake paid work while they hinder others. Research suggests that the options are greater for women with small able-bodied children than for women with aged dependants, and easier to negotiate for women with greater class capacity. There is a definite social policy 'advantage' for women with young dependants that is not available to women with aged or infirm dependants. Of course, accom-panying this general claim must come recognition of the diversity amongst Australian women that complicates the picture – particularly differences of class capacity and location.

By way of conclusion, I wish to identify a number of issues for further research that is necessary to substantiate and amplify my argument.

First, the precise reasons for the life course bias in the social policy set to which I have referred need investigation. Three factors deserve mention as possible starting points. One is the 'femocrat' factor – activist femocrats within the Australian state (themselves not yet 'women in the middle', for the most part) have pushed childcare as quite basic to women's liberation, along with programmes relating to equal employment opportunity. The forms which childcare provision has taken have been unlike the universal, subsidized provision envisaged by many activists, and have favoured middle income, professional, Anglo women. As analysts of the State have commented, the channelling of demands through the State tends to transform those demands. A second possible influence is that women responsible for caring for aged dependants simply have not the time to be activists in the cause of creating employment opportunities for themselves through better services. (After all, they rarely have time even to meet to discuss their experiences of caring, as researchers in this area have discovered (McCallum and Gelfand 1990).) Any lobbying on this matter would be most difficult at present, when there is not the slightest indication that any part of the state apparatus considers this a priority or has started to implement a scheme to support such women. Particular difficulties in confronting this situation, by definition, would be experienced by women with less class capacity. A third possibility is the differences in social expectations of different cohorts of women. Those who are presently 'women in the middle' may have sought (or been restricted to) employment less in conflict with their caring roles; this may be less the case with those younger women whose interests are relatively well represented in contemporary urban social policy.

Second, the life course implications of the spatial policies of the State need specific attention. Much of the contemporary critical literature on the State that addresses issues of gender or of life course, focuses on policies that are social and economic and ignores the explicitly spatial directives of governments. In Australia at present, some major revisions of spatial policy are occurring. The implications of these for women of different life course circumstances are worthy of close attention. These include:

1 Suggestions that policies of 'urban consolidation' are necessary to limit urban expansion and the costs of urban infrastructure.[2] Some local councils, both on the urban fringe and in middle suburbs, have begun to change the sizes of permissible housing blocks, providing for the possibility of dual occupancy of already-occupied blocks and so on. If this produced cheaper housing through the metropolitan area and at the urban fringe, it might have advantages for women of lower income, women without the income support of a male salary, single

older women, single parents and so on. However, evidence to date indicates that urban consolidation is unlikely to increase housing affordability, because it fails to address the causes of housing price inflation (Burke and Hayward 1990: 45).

2 There are suggestions that a series of new towns be constructed, or that regional development in existing provincial towns be facilitated (*The Age* 25 January 1991: 3, 10) – the non-metropolitan settlements in all these proposals would be linked to the metropolitan area by effective public transport. What does this mean for women in different life course circumstances? Country towns are not renowned for their extensive social services; if women seeking affordable housing move to these locations their isolation and difficulties could continue – the stereotypical urban fringe will appear in provincial towns. The establishment of services and physical infrastructure to cope with a new population influx would be expensive and unlikely to appear immediately. It is not clear that women at any life stage would be better situated by the isolation and under-servicing that a decentralized settlement might offer them.

3 The Victorian State Government has coordinated the reshaping of some inner city spaces, helping private investors to plan and produce huge, inner city developments on state-owned land, combining offices, retail space and middle- to luxury-cost apartments. In an interesting twist, these inner urban, luxury (and certainly never low-income), partly-residential developments are seen by policy-makers as one way to increase the overall housing and population density of the metropolitan area, offsetting continued urban sprawl (*The Age* 25 January 1991: 10). What is not recognized, it seems, is that the same income, and possibly life course, groups will not be served by the dense, inner-city developments as have been served by the spreading housing estates on the urban fringe.

Finally, as was evident early in this chapter before the discussion of the Australian empirical material took place, it is important for research to examine life course variations as it addresses the diversity of people's experiences already considered in feminist theory and research practice. In particular, feminist investigation of the State requires systematic attention to life course issues in its analysis of the outcomes of state interventions for different groups of women and its consideration of the participation in activism of women in a range of circumstances.

13

MAKING CONNECTIONS
Space, place and the life course
Cindi Katz and Janice Monk

You know her
She keeps walking

Even at eighty
when her knees ache
and she needs a winter coat in May,
she keeps walking.

Even when her ankles swell,
her bones frail fingernails,
her heart uphill
louder than car horns,
she keeps walking.

She buys thick heels, takes slow steps,
her pale eyes still gauging the safe route.
She pushes her grumbling body on.
She tells it:

we can browse stores whenever we want.
We can buy Listerine, mints, that new hand
cream that removes wrinkles, cologne
if it's on sale.

(Mora 1991)

Pat Mora's *Stubborn Woman* evokes several of the themes our authors have developed as they explored the geographies of women over the life course. Her mobility is clearly affected by her stage of life – at eighty she is no longer agile. As a woman, she keeps an eye out for 'the safe route' while undertaking a shopping trip; that she is walking probably indicates her class and cohort. Yet she perseveres, fully aware that her autonomy and mobility are interconnected, and she maintains the desire of her earlier years for the small luxuries that enhance the image of her body.

In bringing together this collection of essays, we have tried to present cross-cultural perspectives on aspects of women's experiences that have been neglected in both geographic and feminist literatures – the significance of space and place in women's construction of their lives and the importance of changes over the life course to women's geographies. The possibilities for the young girl, for the mother with young children, for the woman seeking new options and bearing changing responsibilities at midlife, for the single woman or the older woman to meet her needs and fulfil her desires are variously constrained and enhanced by the places in which she lives and the spaces she must negotiate. We have introduced a number of concepts that are useful for understanding the life course. Among them are the importance of prior experience; the significance of cohort; the potential for continuing changes in motivations and behaviour; the diversity of roles that women fulfil in the home, the workplace and the community both simultaneously and over the years; and the ways in which ideologies about gender and age affect individual opportunities and structure collective experience. These issues were considered in each chapter, with particular emphasis on the ways in which they articulate with women's geographies. In this concluding chapter we will draw out the geographic implications of these life course concepts to illuminate critical intersections between space and time, place and person in women's lives and the larger social relations with which they are mutually determined.

While the tight weave between production and reproduction in women's lives has perhaps become a commonplace, at least in feminist literature, the spatial parameters and place-bound significance of this relationship in women's everyday lives have not yet been addressed adequately. The question of mobility and its antipode, rootedness, is of great significance here both in terms of daily local patterns of women's movements and longer-term migrations. The intersections between women's activity patterns and the social relations of production and reproduction at various geographic scales are also illuminated in the various chapters.

By using the geographic constructs of mobility and scale to address the meanings and manifestations of sociocultural and political-economic relations over the life courses of women, we can begin to theorize commonalities and key differences in women's experiences over the life course, without becoming mired in their overwhelming diversity through time and across space. As we distil these key differences and analyse common bonds in the geographies of women over the life course, we suggest some of the policy and practical implications of our research.

The metaphor of juggling appears repeatedly in the various chapters. We join the women who are the objects of our authors' inquiries in juggling disparate phenomena – space, life course, place, sociocultural patterns, political economy, biography formation and history to provide new insights for both feminist and geographical analyses of women's experience. In this concluding chapter we will review our findings to summarize how the concepts we have introduced illuminate women's geographies. We turn first to the theme of mobility, then examine the significance of place over the life course, considering local and larger scales. As we consider these themes, we will also reflect on some policy implications of our research, hoping thereby to point to ways of enhancing the quality of women's lives at whatever points in the life course they may be.

MOBILITY AND THE LIFE COURSE

The theme of mobility has been pervasive throughout this book. On a daily basis and over the long term, women must move to secure their economic support, integrate their productive and reproductive activities and take advantage of opportunities for personal development and autonomy. Their capacity to do so is often severely constrained by ideologies about sexuality and appropriate behaviour for women, by the responsibilities of their gender roles and by a variety of political, economic, social and environmental conditions. Within the general patterns, however, are many variations which reflect the context of family and community, circumstances of class and cohort, prior personal experiences and individual motivations. We do not wish to repeat the many examples our authors have provided. Rather, we will highlight a few of the complex ways in which women's mobility is shaped and expressed over the life course. We will first consider aspects of daily or frequent movements, then discuss longer-term changes in place of residence and community.

The daily movements of young girls are important to understand, not only because of their immediate consequences but because they have long-term implications. Girls' freedom to move independently and to encounter directly diverse environments varies considerably across geographic settings, sometimes in ways that differ from commonly held impressions. Thus, as Cindi Katz shows, the pre-pubescent girl in Sudan experiences greater equity in spatial range with boys in her cohort than her counterpart in a large metropolis in the United States, and may have considerable freedom to explore the environment. Several factors affect children's mobility. As both Katz and Lydia Pulsipher suggest, settlement forms which include extended family compounds offer potential for casual surveillance of children by many adults and create an environment of greater security for children. Under these conditions, girls move relatively freely within a circumscribed space, and are not as housebound as many in more isolated living conditions, such as those in urban high-rises. Additionally, if in these societies girls are seen as valuable contributors to productive and reproductive work, rather than primarily as dependants, their mobility may increase to accommodate the accomplishment of their tasks. Whatever its origin, the expansion of spatial range offers enhanced opportunities for environmental learning.

Even within settings where girls experience some freedom, however, the gender division of labour may inhibit girls more than boys. Lydia Pulsipher comments that boys' tasks permit them a greater range beyond the houseyard than is available to girls. The distinction is more marked in the isolated nuclear family farm homes of Colombian frontier settlements described by Janet Townsend, where a dramatic gender division of labour confines young girls and most women to household space. Such consequences raise questions that policy-makers and planners working with frontier development rarely consider, though they need to be addressed if women's long-term well-being is to improve.

Ideologies that link independent mobility and sexuality also affect growing girls in crucial ways. As Katz indicates, fears for girls' safety as well as norms regarding control over pubescent females contribute to the imposition of greater spatial restrictions on girls than boys, though boys may in fact be subject to considerable and greater danger. Still, these restrictions are not immutable. She offers hope but also expresses concern about the possible changes in gendered norms of mobility as economic changes proceed and cultural ecologies are modified.

Within Western urban societies, children's freedom to move varies by geographic context. Jeanne Fagnani notes that the decision of upper-middle-class Parisian families to relocate from the city to the suburbs is often motivated by the parents' desire to give children greater freedom within a secure environment. Though she does not distinguish between boys and girls, Sandra Rosenbloom demonstrates that Dutch children more often take responsibility for their own movements than children of similar ages in the United States, largely because of the extensive use of bicycles in the Netherlands, compared with reliance on automobiles in the United States. She also points out how different familial structures affect children's daily travel. Children in two-parent families have greater mobility than children of single mothers. She notes, however, that this greater mobility does not stem from fathers in dual-parent households assuming more responsibility for transporting children; rather, it is that single and married mothers behave differently. These examples suggest that not only do early experiences of mobility have implications for the development of girls' spatial competence and environmental knowledge that affect them in adulthood, but also that the varying life course experiences of their mothers, such as remaining married or divorcing, may contribute to creating mobility differences among children, especially in settings where other adults are unavailable to provide surveillance or assistance.

The future lives of young girls may also be affected by movements of greater magnitude than the daily ones discussed above. Cindi Katz describes a situation where the migration from rural communities of young men is widening the demands on young women to expand their labour, thereby requiring that they reactivate the environmental knowledge they gained in childhood. By contrast, Janet Townsend documents migration away from rural frontier communities by girls after about the age of ten, because they are seen as surplus consumers and thus as a drain on their families. They go to the city for work as domestic servants if they are from poorer households, or for education if their parents have more resources. Such migration clearly affects later life chances and, as Townsend notes, also has implications for mothers; they may move temporarily to be with their school-age daughters or, in later life, take advantage of their children's urban connections as a way of escaping from the heavy demands of household work in the predominantly male frontier communities. Her chapter also helps us to see the importance of life course timing to the experience of migration.

Most of the differences in mobility among girls that we have discussed reflect the specificities of the places in which they are growing up. When we examine the geographies of women, the picture becomes more complex as the effects of different prior choices and earlier experiences accumulate. Popular opinion, public policies, private businesses and scholarly work have made much of the ways in which women's options are constrained as a result of their child-bearing and -rearing roles. Though the essays in this book support this interpretation, they also go beyond the idea that women are bound in some permanent and homogeneous manner by being mothers; indeed their discussions of the mobility of women in early and middle adulthood clearly support the position that women shape their own lives in ways that increase the differences among them.

Sandra Rosenbloom's study of daily travel documents that employed married women in the United States bear more responsibility for transporting their children than do their husbands and also that these women are more likely than their husbands to link their journeys to work with trips for other purposes. Her detailed analyses, however, reveal that the extent of this travel and the degree to which it differs from men's shift as children grow up, with fathers' contributions diminishing over time. For example, the number of married mothers' linked trips is greatest when their children are under six years of age, lessens when the children are between six and twelve years, and increases again when children are in their teens. Single employed mothers, however, are less likely than married mothers to make either linked trips or trips solely for their children, though the distinctions between the two groups of women are least when the children are young. As Rosenbloom points out, her data present patterns of difference, not their causes, yet she suggests that single mothers may limit deviations from direct journeys from home to work because no one else is available to stay with the child, or because they cannot afford the costs of longer childcare. Her findings regarding the different patterns of single and married mothers and the related consequences for their children provide a new challenge for those attempting to develop new models of adequate and affordable childcare.

Widowed and divorced women interviewed by Geraldine Pratt and Susan Hanson also differed from married women in their attitudes towards the journey to work. Though not in fact employed significantly further from home than the married women, they expressed greater willingness to travel further to work than their married peers. Differences in daily mobility among women originate not only from the demographic aspects of marital and family status, however. Rather, Kathleen Christensen found that the women who chose to work at home were implementing value choices about family and work life. For the younger of these women, the stresses of inadequate daily transportation or the demands of a job that required out-of-town travel were incompatible with their notions of motherhood. Unwilling or financially unable to forgo employment, they accommodated by developing businesses at home. They differ from the midlife women working at home whose choices reveal not only their life stage but also the significance of generation in experience. The older women had spent many of their child-rearing years outside the paid workforce, reflecting the values and economic conditions of their cohort. Re-entry into satisfying employment proved difficult, shaping their decision to work at home. The older women's patterns of mobility related more to their husbands than to their children. One, for instance, organized her work schedule to fit in vacation travel with her husband; for another, the satisfactions of her home-based work, which required travel, contributed to marital strains that resulted in divorce. Many of the mature married women Pratt and Hanson studied held jobs close to home, reflecting their cohort's greater concern for home and family over the job. Still, these women were less likely than younger women to say that they chose jobs because they were locationally convenient to childcare or schools, or had flexible hours that fitted with children's needs. They were more likely to describe choosing their jobs because they were in the fields in which they wanted to work or for which they had been trained. Both chapters remind us of some of the costs to women and to society of attitudes that place greater responsibility on women than men for reproductive work, and they show us how employers' practices reinforce such social norms.

Class and cohort variables intersect with values to shape the decisions of the Parisian women Jeanne Fagnani describes. Well educated and wanting to sustain careers, they represent a generation influenced by the contemporary women's movement. They incorporate both new values for shared parenting with old values about the responsibilities of the good mother. To meet their own desires as well as those of their families, they take advantage of their class position and dual-earner status and sometimes consider shifting their place of residence to accommodate their goals. For some, the preference for a suburban lifestyle prevails; others choose to live in Paris. The decision reflects class capacity, cohort, personal values and anticipation of their future needs, as well as their present life stage.

Various structural conditions influence the travel and residential patterns of women with children of pre-school and early-school age, as Damaris Rose indicates. In the Montréal setting she describes, policies on school bus pick-ups, for example, are tied to age of the child as well as distance from the school. In order to secure transportation for their children, parents often must decide whether to change their residence or workplace, or else juggle their employment schedules in complicated ways to fit their children's lives. Alternatively, they may call on older children or other family members to provide services. Rose shows how the latter possibility is more likely in certain ethnic communities, reflecting not only the cultural values of these groups but also the ways in which Canadian immigration policies have structured the demography of different ethnic groups.

Class and life stage variables also affect the mobility of the Sudanese and Colombian rural women studied by Cindi Katz and Janet Townsend. Intrusion of capitalist agriculture into Sudan has created new demands on families to generate cash income and to increase their work loads. Though adult women from relatively well-off households retain the restricted mobility of purdah, those beyond the earliest child-bearing years from most other households work in the fields to increase their family income. These women also go out to fetch wood and water, moving in the company of other women. In Colombia, it is the women from the more affluent community of El Distrito who have the option to escape the isolation of the frontier settlements by establishing temporary second residences in town with their school-age children. Those who are poorer usually must remain, to meet the demands of domestic work in these predominantly male settlements. Single women, however, have little option other than to leave permanently, even when they are relatively skilled teachers or bookkeepers, since they are not valued as permanent contributors to the community unless fulfilling the role of housewife.

The chapters dealing with older women quite clearly demonstrate how cohort and place of residence affect their mobility. As Sandra Rosenbloom shows, in the United States where the car is essential in so many places, whether a woman holds a driver's licence is central to her mobility. The likelihood of holding a licence is related to cohort, with women over seventy significantly less likely to have held a driver's licence at any time in their lives than women now in their fifties. Indeed, though older women's mobility is related to their marital status, it is the woman's holding of a driver's licence, not her marital status, that makes the most impact on trip frequency. Rosenbloom's research provides a necessary reminder to transportation planners that long-term needs may differ from those they currently identify.

In the Colombian and Caribbean contexts we have seen how older women's longer-term moves reflect the life courses of their children or other significant kin. Whether older Colombian women have the option of more comfortable lives in urban areas, or must remain on the frontier, depends on the life courses of their children. Among the elderly women of Montserrat described by Lydia Pulsipher, some have moved to town where they are supported by the remittances of kin who have themselves left the island. Janet Momsen also highlights the significance of remittances from immigrant relatives in the support of older women in the Eastern Caribbean, but notes that declining rates of labour force participation within the region by women over the age of forty-five are thought to be related to emigration by older women joining adult children who have left the Caribbean. Among the Montréal grandmothers providing childcare to the families Damaris Rose studied are those who have come to Canada under the auspices of immigration policies which have favoured family reunion for certain groups.

Many of our interpretations have indicated that mobility provides greater autonomy for women, though we have also shown how women's geographies are tied to the constraints of their gender roles and, in particular, to the demands of their reproductive work. But the inverse of mobility is not always immobility; there are positive polarities as well. For example, rootedness offers many women personal satisfactions and rewards, as well as possibilities for social life and the sharing of burdens in productive and reproductive spheres. The case is made most clearly by Patricia Sachs in her history of a former mining community in Appalachia. The sharing of lifelong experiences of struggle and of a particular lifestyle has bound people to place and one another. They enjoy a social support that is missing for those who move to unfamiliar places, especially without close kin. Sachs reports that some families have returned to their Appalachian homes, despite the economic limitations they face there. Rootedness within the houseyard, an environment where she has held power, also provides a secure niche for the older Caribbean woman even as others move away. As Lydia Pulsipher argues, the flexibility of Caribbean gender relations makes it possible for some women to sustain themselves in old age. Finally, we note the loss of social support experienced by Colombian women on the frontier, away from home communities where they could call on established networks of family and friends. On the frontier, in predominantly male communities, women may be so isolated that they even give birth alone. Janet Townsend offers insights on how this isolation is compounded by women's confinement to the spaces of their homes.

SCALE AND THE LIFE COURSE

The last examples, focused on the body, the household and the community, suggest how the construction of scale in women's everyday lives affects and is affected by their social relations. Using the geographic concept of scale, it may be possible to draw out pervasive themes and to connect the most salient theoretical and practical issues at the heart of women's geographies over the life course. Feminist geography has demonstrated and called attention to the ways that the production of space reflects, reinforces and reproduces unequal gender relations (Bowlby *et al*. 1982). Although they vary in form, these processes work at all scales, and the ways in which they come together in particular times and places have implications for women across the life course.

Cindi Katz discusses ways in which the construction of fear and danger at the scale of the body limits girls' movements through the local environment more than boys', restricting the girls' spatial range. This, she suggests, affects the development of their spatial skills and their acquisition of environmental knowledge; further, the restrictions on their autonomous movements are of consequence for girls' psychosocial development, as they learn to fear being alone in particular places. Katz's research complements that of Gill Valentine (1989) and Rachel Pain (1991) who argue that ideological constructions of fear operate to exclude women from the public sphere and mystify the fact that more crimes against women's bodies occur in the private realm of the home or the workplace. Katz demonstrates how these constructions operate in different settings to limit girls' freedom as well. She points to the need for urban social policy to recognize that most children no longer have stay-at-home caretakers, and to address means of ensuring children's safe access to the outdoor environment.

The organization of space at the scale of the household reflects the articulation between the relations of production and reproduction. Feminist theorists have long analysed the close association of production and reproduction in women's lives, noting women's enduring responsibilities for the bulk of the 'caring' work of reproduction in most parts of the world. This responsibility obtains, except in rare circumstances, even under conditions in which women participate fully in the work of production. The increasing separation of the home and workplace with the rise of capitalism and industrialization in the West marked a significant, and highly spatialized, disjuncture in women's and men's relations to production and reproduction, and to one another. A particular social contract defined and reinforced this relationship, which supported both the accumulation of capital and the maintenance of patriarchy. This social contract has been rendered obsolete in the United States and other areas in the industrialized West since the 1950s by the failure of the 'family wage' to support most families and by the women's movement which changed ideologies about women's lives and expectations. Both phenomena have contributed to sharp increases in women's participation in the waged labour force in the industrialized West.

This participation in the labour force has not been matched by significant declines in women's share of the work of reproduction, however. One spatial means which they have adopted to accommodate their dual roles is to undertake paid home-based work, a practice that has grown dramatically over the last two decades in the United States and elsewhere in the industrialized world, and which is being fostered in many other regions (cf. Beneria and Holdan 1987). Kathleen Christensen points out some of the contradictory outcomes of this increasingly popular individual strategy, and highlights the mixture of gains and losses that accrue to the people involved, their families and the larger society. In addition to discussing how the space of the home itself is altered to facilitate paid work, Christensen demonstrates the varying effects of this arrangement for women at different stages of the life course. She notes how home-based work at once perpetuates women's more fragile connection to the benefits of paid employment, encourages the incorporation of unpaid labour, whether by women themselves or by members of their families, and can reinforce traditional sex roles. To this end she notes the importance of creative benefit and pension packages that recognize home-based workers' contributions to the economy, and remove them from the netherworld between production and reproduction, to which they are often relegated. Christensen also makes clear that as individuals and communities cope with the local experience of the restructuring of global capital and changing definitions of the family, bringing production and reproduction together within the home may be one of the most attractive solutions to a wide range of women and families, despite the problems she documents.

The work of Patricia Sachs points to another arena in which global or regional economic problems are experienced and addressed at the scale of the home. Sachs focuses on a community in West Virginia reconstituting itself in the wake of mine closure. She documents how women and men elaborated their home-based and home-oriented work to secure a meaningful life for themselves after losing their community's main source of income and definition. Her research revealed that, for older residents who had known no other work but mining, it was more desirable to stay and eke out a modest living rooted around their homes and gardens than to move to locations where they might have opportunities to earn larger incomes. Sachs's work demonstrates how a community can be steeped in a particular way of living developed over a lifetime, lose that definition, and yet abide through the highly local efforts of its residents. These efforts did not just enable the people to secure some sort of livelihood, rather they simultaneously changed the pattern of social relations between men and women within the home and physically improved the community as well. They were enabled, however, by the residents' ability to call on public financial assistance. For younger people, these solutions were not viable and they were forced to relocate to secure employment.

The neighbourly cooperation Patricia Sachs documents is mirrored in the mutual aid relationships of the houseyards noted by Lydia Pulsipher and Janet Henshall Momsen in the Caribbean and Cindi Katz in Sudan. Many of the tasks of reproduction, particularly those associated with child-rearing, are shared within these collective family spaces. Such arrangements, largely short-term in Sudan but often lasting years in the West Indies, enable women and children to participate more easily in economic activities outside the home. Katz suggests that houseyards also offer children a secure base from which both to engage in multiple social interactions with peers and elders and to explore the physical environment. Pulsipher and Henshall Momsen note how the sociospatial arrangement of the houseyard both suits and facilitates the relative autonomy of women in the Caribbean who frequently have children out of wedlock and with multiple partners. The mutual assistance of relatives and friends, both men and women, centred largely on the houseyards, fosters reciprocal means of support across the life course, whether this be by means such as farming to provide for the family, sending remittances or caring for children, sick people and older kin.

Beyond the gender relations of production and reproduction, several chapters make clear that global capitalism acts locally, and the 'geography of the book' binds some of these local manifestations. The very displacements discussed by Janet Henshall Momsen, Lydia Pulsipher and Janet Townsend in South America and the Caribbean produce a labour force for the industrialized economies to the north. The immigration policies of Canada described by Damaris Rose, for example, encourage women from the former areas to emigrate alone to the latter to serve as domestic labourers. Thus they assist middle-class women caught in the squeeze of their productive and reproductive roles, but able to afford a private solution. The affordability of the solution is, of course, predicated on the low wages of the immigrant women. In turn, their ability to survive on these wages is subsidized either by the 'free' labour of relatives in their new homes or by the existence of an extended family structure centred around the houseyards 'back home' which cares for their offspring or other dependants. These relationships are often inter-generational and thus represent a spatial pattern that fuses life course, culture, class relations and the international political economy. Immigration policies of various industrialized nations draw on or thwart these relations in some places and for certain groups. For example, Australian or Canadian policies have supported family reunion as a route to immigration by Europeans, but Canadian policies have encouraged lone women to immigrate from the Caribbean for low-paid work.

Ruth Fincher discusses some of the disquieting aspects of this relationship between women across social formations in her chapter on Melbourne, Australia as well. She focuses on the as yet little addressed conflicts in middle-aged women's lives between the demands of caring for children and the needs of others such as aged or infirm relatives. A peculiarity of the juxtaposition of two life course characteristics in the industrialized West – the postponement of child-bearing and wide-spread increases in life expectancy, along with the stubborn relegation of caring work to women – often results in women in their middle years juggling not only paid work and child-rearing, but also the care of their older family members. The dispersal of families across cities, regions and even nations make these intersections even more difficult to negotiate, as Fincher's research makes clear.

These extensive webs of care which absorb women's time, as well as emotional and physical energy are made simpler or more difficult to deal with depending upon their particular spatial form. Difficulties at the intra-urban scale are addressed by Damaris Rose writing on Montréal, Québec and by Geraldine Pratt and Susan Hanson in their chapter on Worcester, Massachusetts, and at the scale of the metropolitan region by Jeanne Fagnani writing on Paris and its suburbs. Rose, for example, discusses the challenge presented by disparate locations of childcare facilities, schools, home and the workplace for employed mothers and children of different ages. She notes that these difficulties generally fall hardest on the working-class and immigrant women who often live in neighbourhoods poorly served by extant facilities and transportation networks, and who frequently have inflexible work schedules.

Rose's work again points to the ways in which women remain responsible for reproductive tasks even under new socio-economic patterns in which they are full-time earners or heads of households. Sandra Rosenbloom underscores this issue in her discussion of women's intra-urban travel patterns. Here again, it is more commonly women than men who are obliged to combine their dual responsibilities as parent and workers, through linked journeys. She notes that public transportation generally best serves the uncomplicated journey to work and is inadequate to meet the needs of parents who must make multiple stops between home and work. Geraldine Pratt and Susan Hanson make a similar point in their chapter, noting that for married women, employment options are constrained by distance from home much more frequently than men's. Their research demonstrates clearly the links between the women's concerns at particular stages of the life course and their employment options and choices within a particular region. Women less geographically bound in their employment decisions tended to have fewer domestic responsibilities, either because they were single and/or had no children, or because they were no longer actively engaged in child-rearing. Their research revealed, dismally, that even in the contemporary urban landscape of the United States, it is rare for women to combine motherhood with a well-paying job. However, even among highly educated, middle-aged, upper-middle-class women, Jeanne Fagnani found in the Île-de-France region that, in the main, women worked closer to home than men. Her findings complement those of Pratt and Hanson, and reflect again the spatial aspects of women's juggling of their dual roles. Interestingly, Fagnani found that among couples committed to sharing equally the work of child-rearing, fully 25 per cent chose to live where their commutes would be equal – a spatial manifestation of changed social relations. Addressing the special problems of 'women in the middle' Ruth Fincher illuminates how the State's and other institutions' privileging of the construction of women at a particular life course stage – 'dependent wife and mother of young children' – can relatively disadvantage women who are outside this category. She also carefully demonstrates how women's class and ethnic characteristics affect their capacity to avail themselves of existing social services, whatever their life course stage.

While our attention to the life course teaches us that women's needs and experiences are different at various times during their lives, many of the chapters reveal how pivotal the relations of production and reproduction are, both in women's lives and in the sociospatial formations of the society itself. At different scales and in vastly different settings, the spatial reflection of the greater parity in child-rearing Jeanne Fagnani detected among heterosexual couples around Paris, and Lydia Pulsipher's example of the Caribbean houseyard with its intergenerational mutual aid strategies which loosen the bonds of patriarchy, each suggests how redefined patterns of social relations may lead to and be sustained by new spatial forms. Women's agency in negotiating and altering these articulations within and against historically and geographically specific structural constraints is the focus of many of the chapters.

CONCLUSION

What is striking in the chapters is how international, state and corporate policy and practice ricochet around women's everyday lives. One of the key lessons of feminist scholarship and the political engagements of women's movements world-wide is that these larger structures are set in place and deployed locally and regionally by historical agents in their everyday lives. In this interpenetration, which takes place at all scales, the structures and practices of the social relations of production and reproduction are maintained, altered, subverted, reconstituted. The spatial forms and meanings of these social relations have been illustrated in each chapter, highlighting different ways they come together and thereby pointing to significant conjunctures between women's roles, the life course and the production of space.

The chapters stand as testimony to the diversity of women's experience under these common bonds, and to the place-specific possibilities of various mediations of particular conjunctures ranging from reciprocal or unpaid (intra-household, extended family) solutions, to fee-based resolutions or state-sponsored interventions. It is important not to get lost in the kaleidoscope of diversity, or sunk in the mire of homogeneity if we are to understand the significant issues at the heart of women's geographies over the life course. The appreciation of diversity often devolves into an over-reliance on individual solutions to collective social problems; conversely, seeing only the similarities in women's experiences can lead to overgeneralized and inappropriate policy decisions. The tendency to position women as dependants and to tether them to certain life stages, notably young motherhood, leads, for example, to policy and practice focused only around this condition – at once naturalizing and reinforcing women's often contingent relationship to paid employment. This constricted focus results in inattention to policies and practices that would facilitate women's full participation in the labour force, no matter what their stage of life. At the same time, this dominant view undermines its own rhetoric by the near total failure of social policy and practice to reward the caring work that women do throughout their lives.

In the absence of such efforts, the structural determinations of capitalism and patriarchy which impinge on women's lives not only remain untransformed, but often are personalized, leaving women to cope individually with their multiple commitments. In calling attention to the unified nature of production and reproduction in everyday life, examining the ways this unity shifts over the life course and varies according to class, ethnicity and gender, and addressing the ways space both affects and is an outcome of this relationship, it has been our intent to show the multiplicity and changing nature of women's responses to similar structural constraints over their life courses. As we construct and

277

reconstruct our lives in the face of material social practices such as modernization, global economic restructuring and migration, we recognize that in the full circles we travel, individually and collectively, in space and time over the near and far, short and long, we not only produce our lives but large-scale processes as well. Discovering the common bonds that tie us, and the rich differences between us as women of different ages, classes, races and ethnicities in specific historical and geographic circumstances may suffuse some circles with greater meaning and new potential, and rupture others to shape new trajectories in the social production of self, space and society.

NOTES

1 WHEN IN THE WORLD ARE WOMEN?

1 The report was prepared for the Federation for Policy on Ageing in the Netherlands and the Age Concern Institute for Gerontology in the United Kingdom by Marianne Coopmans, Anne Harrop and Marijke Hermans-Huiskes. Its major findings are summarized in *Network News: Global Link for Older and Midlife Women* 5 (1) Spring/Summer, 1990, pp. 9–15.

2 Geographic studies of children have been reviewed by Hart (1983); studies of older women are reviewed in Zelinsky, Monk and Hanson (1982). Little has been added since that time. Though a small number of geographers continue to conduct research on topics related to ageing, this work often does not identify gender concerns, nor address feminist questions. Recent geographic research in the United States on ageing is summarized by Golant, Rowles and Meyer (1989).

3 See White *et al.* (1989) for a summary of recent work.

2 WOMEN AND WORK ACROSS THE LIFE COURSE

We wish to acknowledge the support of the National Science Foundation (SES-8520159; SES-8722383) and the National Geographic Society. We are grateful to these funding sources. We also thank Melissa Gilbert and Glen Elder for their research assistance.

1 Of course much of women's waged employment has exactly these characteristics but, as Pringle (1989) has pointed out in relation to Australian secretaries and Phillips (1987) with respect to working- and middle-class women in Britain, there are some very important class distinctions within female-dominated occupations that require more careful consideration, not the least to explore the tendency for downward mobility throughout the life course.

2 Two qualifications must be considered when interpreting this comparison: first, the statistic from 1950 includes households without children, and second, the 1985 statistic understates the percentage of households which pass through this stage at one point in their life course.

3 A recent court case (*California Federal Savings* v. *Guerra*), for example, let stand a California statute that entitled female employees to four months' leave without pay for a physician-certified disability resulting from pregnancy or childbirth. An employer had challenged the statute on the grounds that it did not provide equal protection for men, but the court of appeal

held that the statute did not conflict with the US Civil Rights Act of 1964. In this instance the court clearly upheld the notion of inherent differences between women and men, differences that warrant different treatment in the workplace.

4 Households were drawn from an area-based, spatially stratified sample of census blocks selected at random from each census tract so as to constitute a representative sample of the Worcester-area working-age (twenty-one to sixty-five years) population. This strategy resulted in an area-based PPP (probability proportionate to population) sample. The interviews averaged 75 minutes. The results reported here are based on responses from 526 currently employed people (190 men and 336 women) in the sample. Within the random sample of households, more women than men were purposefully interviewed to permit disaggregation of the female sample into the three gender-based occupation types.

5 Bird and West (1987) report on British studies that document the downward occupational mobility experienced by women upon marriage. Treiman (1985) reports US cross-sectional studies that document the (negative) effects of marriage and children on occupational status.

6 The divorced, widowed and separated women were quite similar to the married women with respect to a number of social characteristics, making the discrepancy in employment experience all the more interesting. Their ages were fairly similar: an average of 44.1 for the divorced, widowed and separated women and 40.1 for the married ones. They had similar numbers of children living at home. Among married women, 26 per cent had no children, and 53 per cent had one to two children living at home. Considering the divorced, separated and widowed women, 23 per cent reported no children at home, and another 57.1 per cent had from one to two children living with them. Their educations were roughly comparable, although there was a slight tendency for married women to be better educated. This is an interesting direction of difference, given that married women are more likely to find employment in female-dominated occupations. Considering married women: 8 per cent had less than high school qualifications, 38.4 per cent had a high school diploma, 5.8 per cent had attended a technical or secretarial programme, 38.9 per cent had some college or a college degree and 7.5 per cent had some post-graduate college education. Among those divorced, separated or widowed: 17.2 per cent had no high school diploma, 36.2 per cent had only a high school diploma, a further 7.2 per cent had attended a technical or secretarial programme, 33.2 per cent had attended college and 5.7 per cent had some postgraduate college education.

7 Not surprisingly this does not translate into higher family incomes among households headed by women. Only 3.3 per cent of married women lived in households with incomes lower than $9,000 while fully 24.2 per cent of divorced, separated and widowed households did. Alternatively, 31.1 per cent of married women lived in households with incomes of $35,000 or more while only 15.1 per cent of divorced, separated and widowed women had this level of family income.

8 We asked interviewees to indicate whether they felt a particular job attribute was very important, important, neither important nor unimportant, unimportant or very unimportant. Fifty-seven per cent of divorced, separated and widowed women felt that 'possibilities for advancement' were important or very important. The comparable percentage for never married

and married women were 62 per cent and 34 per cent, respectively ($X^2 =$ 16.3; d.f. = 4, p = 0.002). For the attribute of 'good benefits', a similar proportion of divorced, separated and widowed, and never married women (53 per cent and 54 per cent, respectively) felt this was important or very important, but only 39 per cent of married women felt this way ($X^2 = 12.72$, d.f. = 4, p = 0.01).

9 We interviewed the owner, chief executive officer or, in the case of large firms, the human resources officer.

10 We asked our interviewers to assess the validity of the employers' 'guess-timates' concerning the family circumstances of their employees. In the majority (60 per cent) of cases the interviewers thought that the 'guessti-mates' were equally valid or invalid for female and male employees. For 29 per cent of the firms, however, the assessment of family circumstances of female employees was judged to be more accurate. For only 11 per cent of the firms was the assessment of marital status of male employees judged more accurate ($X^2 = 119.98$, d.f. = 16, p = 0.000).

11 Divorced (and separated and widowed) women were willing to travel an average of 30.1 minutes to paid employment (s.d. = 13.1) while married women were willing to travel only 26.5 minutes from home (s.d. = 13.6, F = 2.3, p = 0.13).

12 This difference is statistically significant: $X^2 = 5.35$, p < 0.10.

13 Nearly three-quarters (73 per cent) of the mature group, versus only half (51 per cent) of the other married women, work full-time ($X^2 = 12.8$, df = 1, p < 0.001).

14 This difference is statistically significant: $X^2 = 4.7$, df = 1, p < 0.02.

15 Only about one-third (34 per cent) of the mature married women versus about half (47 per cent) of the other married women said that there was another job title at their current workplace that they would see as a promotion ($X^2 = 4.3$, df = 1, p = 0.04).

4 GROWING GIRLS/CLOSING CIRCLES

I am grateful to Sallie Marston, Janice Monk and Neil Smith who commented on an earlier draft of this chapter; and to Harouna Ba who provided research assistance. The research on Sudan was funded partially by a pre-doctoral grant from the National Science Foundation.

1 Much recent work by feminist and post-colonialist theorists has addressed the over-determination of Western notions of purdah and the misrecogni-tion of the agency and strategic intent of women who adopt 'the veil' for their own ends (for example, cf. Fernea and Bezirgan 1977; Lazreg 1988; Swedenburg 1989).

2 With inclusion in the Suki Agricultural Project, many of the goods that had been freely available (e.g., wood, land) or shared commonly (e.g., meat) gradually became commodities. In the case of meat and other foods, this transition – which was underway before the Project – was completed by 1981. At that time, deforestation and the labour demands associated with the cultivation in the Project, were leading to the gradual commodification of wood. Participation in the Project also had exacerbated existing socio-economic inequalities in Howa, and led to new levels of differentiation between village households of tenants and non-tenants alike. With these changes, labour power became a commodity as well, bought and sold among

281

villagers to replace gradually the communal work parties (*nefir*) that had prevailed before the Project.

3 Because it was nearly impossible for tenant farmers to break even, cultivating cotton in the Project (due to high production costs which were deducted directly from gross receipts on the yield prior to paying the tenant; low yields caused at least partially by the lack of farm machinery and timely applications of agricultural inputs; and because cotton was marketed through the state marketing board at a fixed price), it was not uncommon for tenant resistance to be played out around the cotton crop, and for all labour allocations to favour groundnut cultivation from which tenants stood to profit substantially in good years. One of the most common and productive forms of resistance was to let livestock graze on the high-quality fodder provided by cotton.

4 Prior to the agricultural project, access to land was by customary right. As boys matured and established their own households they either took over some of the parcels of their fathers, or cleared new land further afield. By the time the Project was established in 1971, the furthest fields were no more than an hour's walk away.

5 The government primary school in Howa, which opened in the early 1960s, was open to boys and girls. A few girls attended in the early grades but none had ever made it all the way to graduation. Coeducation was frowned upon by most of the local population and only 4.2 per cent of the school-aged girls in Howa attended. The construction of a girls' school, then, reflects a new commitment to female education in the village. This was one of many practical responses to the rapidly changing socio-economic conditions there.

6 Cf. Adolfo Mascarenhas (1977), who discusses the futility of constructing schools without a parallel commitment to providing piped water supplies, in rural Tanzania.

7 Evidence suggests that the outdoor activities of children in high-rise dwellings are restricted because it is more difficult for them to communicate or be in visual contact with a parent or other care-giver than children in low-rise buildings. Even within high-rises there is an ecology of children's experiences – Bjorklid (1985) notes that children who live on upper floors go out even less often than their peers on lower floors. Children who live in high-rises have a higher incidence of respiratory ailments than other children, the apparent result of their having inadequate time outdoors (Pollowy 1973, cited in Bjorklid 1985: 96).

8 Harris (1981) cites numerous studies that indicate a correlation between certain sex hormones and neurologic propensities. In human neonates, for example, there appear to be correlations between female hormones and increased sensitivity to aural stimulation and between male hormones and increased sensitivity to visual stimulation. The former is assumed to be associated with female superiority in linguistic ability and the latter with male superiority in spatial ability. As with all physiological explanations, these correlations suggest propensities only. It is through social interaction and practice that the extent, nature and meaning of their development is determined.

5 'HE WON'T LET SHE STRETCH SHE FOOT'

1 I use the terms 'houseyard' and 'yard' interchangeably to mean associated dwellings and the specific space surrounding them.

2 Studies that establish the ubiquity of the African American houseyard include Pulsipher and Wells-Bowie's geography/architecture study comparing yards in Daufuskie, South Carolina and Montserrat, West Indies (1989); anthropologist Jopling's typology of Puerto Rican houses (1988); home-economists Roberts and Stefani's study of Puerto Rican families (1949); Westmacott's study of gardens and yards in Georgia (1990); and unpublished field notes from Louisiana, Texas, Illinois, Virginia and in the state of Bahia in Brazil (Pulsipher, 1970–87).

3 'Chattel' refers to movable property. In the Caribbean, especially in the post-emancipation period when land ownership was practically impossible for most poor agricultural workers, they often invested in small wooden houses that could be moved easily should they lose the right to live on a given plantation.

4 West Indians tend to figure kinship through the female line and live in matrilocal patterns, hence half-siblings through the maternal (uterine) line tend to live together and to be much closer than half-siblings through the male line. In fact, Smith (1988) reports that uterine half-siblings resist the idea of being designated 'half'-siblings.

5 The custom in the Caribbean of 'borrowing' children is old. Slaves frequently raised children they had not borne, sometimes in order to take care of the needs of the orphan; but more often the child's mother freely gave it to a childless sister or friend so the adoptive parent could have the companionship and loyalty of a child as it grew to adulthood. The custom survives today.

6 'Outside' children are those born either before marriage or present liaison, or those parented by a mate other than one's present mate. Women rarely have outside children while married or cohabiting; but men often do.

7 The research design of many studies masks the rarity of such family units by assuming that couples designated (either in statistical compilations or in interviews) as married were legally joined early in their lives and had all their children together in wedlock. Careful review case-by-case has proven that this is hardly ever the situation.

8 This research was supported by a grant of volunteer research assistants and funds from Earthwatch/The Centre for Field Research, Watertown, MA.

9 While the purpose of this study is not to draw connections between the Caribbean and Africa, there are fertile possibilities for future work. V. W. Turner in his study of village life in the Belgian Congo, explains that due to maternal descent rules there, the role of men as brothers of women and uncles of their children is more enduring and more crucial to attaining power and status in the village than their roles as husbands and fathers. In fact, the children of a headman have a natural rivalry with his nieces and nephews, whom he may regard as closer kin than his own children (Turner 1957). For further comments on the roles of Caribbean uncles see Pulsipher 1992.

7 GENDER AND THE LIFE COURSE ON THE FRONTIERS OF SETTLEMENT IN COLOMBIA

Thanks are due to the Associacion Campesina de Colombia, to HIMAT (the engineering section of the Colombian Ministry of Agriculture), to the Co-operativa Azul, Bolivar, and especially to the Comuna Integral de Payoa, and to all those who cooperated in the surveys.

1 Such a pattern has also been reported in Sri Lanka, with more men than women moving in (the demographers' 'male pioneer model'), and with male predominance being maintained by more women than men moving out (Kearney and Miller 1983).

8 OLD TIES

1 The residents of Deckers Creek in the 1970s had not all lived in the community for the same period of time. One woman had been born there in 1906 and had never lived anywhere else, while others had lived in similar company towns within a twenty-five-mile radius. Some of my data describe other similar towns in the 1920s and 1930s.
2 The women were between the ages of sixty-five and ninety, with most in their seventies. Gerontologists have defined several categories of old age: 'young old', 'old' and 'old old', to account for generational differences among people over age sixty-five.
3 All the information regarding the population size, percentage of the aged population and systems of exchange refer to 1978 and 1979 when I lived in the community.
4 The generational differences in Deckers Creek were matched by occupational differences: the elderly were retired coal-mining families, the middle-aged people were local business people or factory workers and the younger people were professional nurses, doctors and engineers.
5 The term 'social formation' suggests that any community or locality is formed by a relationship between historical epoch and local conditions. It suggests that the social life of communities is constructed not passively, but actively, and will change with shifts in local economy and politics.
6 In the late 1970s, every elderly family owned a kerosene stove to provide extra heat during winter months, even though electric and gas heat was available, and ample, at the time.
7 In the first half of the century, women did not take in work for money. This occurred only when the mine closed and the company no longer owned the community.
8 The awareness of the workers about their work conditions was expressed through their participation in organizing the United Mine Workers Union (UMW). This was vehemently opposed by the company. For a year or so during the 1920s, a rifle tower was set up at one end of Deckers Creek, and anyone who visited a neighbour soon received a knock at the door and was questioned about the purpose of the visit. One man (rumoured to be a union organizer) was shot from the rifle tower, while walking down the street in Deckers Creek.
9 If a family was caught buying any goods from a store in a nearby town, the man would lose his job and the family its house. This extraordinarily difficult situation was the subject of the song whose lyrics read: 'Sixteen tons, and what do you get? One day older and deeper in debt. St Peter don't you call me, 'cause I can't go, I owe my soul to the company store.'
10 From a transcript of taped discussion with one of the elderly women of Deckers Creek.

9 LIFE COURSE AND SPACE

This research was funded by a grant from the Caisse Nationale des Allocations Familiales. I would like to thank G. Desplanques, from the INSEE, for his help.

1 For dwellings bought between 1979 and 1984 inside Paris, one square metre cost US$1,411 (8,466 FF) compared with US$952 (5,710 FF) in the Greater Paris suburbs (Source: Housing Inquiry 1984, INSEE).
2 Between 1962 and 1982, high-level executives and independent professionals increased by 40 per cent in Paris, whereas blue-collar workers decreased by 44.8 per cent (Massot 1990).
3 Seventy-four per cent of the regional female job opportunities are located in Paris and in the inner suburbs; that is 42.4 per cent in Paris proper and 32 per cent in the suburbs. Male employment is more spatially dispersed.
4 These perceptions of the outlying suburbs are historically rooted. A. Faure (1990) demonstrated that, as far back as the end of the nineteenth century, these areas were subjectively associated with 'nature' and 'countryside' for all Parisians, irrespective of their social milieu.
5 In fact, they are reluctant to give up family work and want this responsibility, which corroborates Weiss's (1987) study results and Thompson and Walker's (1989) observations.
6 In 1964, the French government undertook to build five 'new towns' in the Île-de-France, a programme which was supported by a network of legal, technological and financial arrangements. They are located at a distance varying between 13 and 35 km from the centre of Paris. The aim of these new towns was to channel residential growth and employment by creating 'restructuring poles' as important local employment and service centres (Cahiers de l'Institut d'Aménagement et d'Urbanisme de la région Île-de-France 1989).

10 LOCAL CHILDCARE STRATEGIES IN MONTRÉAL, QUÉBEC

This chapter was prepared while the author was Visiting Scholar in the Department of City and Regional Planning at Cornell University, during sabbatical leave from INRS-Urbanisation. Some of the material was first published in a different form in *Political Geography Quarterly* 9 (4), 1990. The research was partially funded by the Social Sciences and Humanities Research Council of Canada (grant no. 410-86-0688). My thanks to the editors of the present volume and to all those who made helpful suggestions on earlier versions; and most of all to my co-researchers, Bernadette Blanc, Nathalie Chicoine, Francine Dansereau and Annick Germain, for contributing ideas that became integral to my thinking. The knowledge acquired during our research is a collective product. I alone, however, take responsibility for the chapter's shortcomings.

1 Although interest in employer-sponsored daycare is increasing, it amounts to only 3 or 4 per cent of total Canadian daycare, and fees are usually as high as in other types of centre (Mayfield 1990).
2 One could be forgiven for suspecting that some of the advocates of 'family policies' want parents, and especially women, to take on the burden of increased 'flexibility', regardless of the effects on their time, their prospects for career advancement or their well-being (*Canadian Dimension* 1989; see Ouellet 1989).
3 This conceptualization springs from a critique of the 'collective consumption' approach to service provision developed by the French school of Marxist urban sociology. For details, see Rose (1990).
4 This said, however, most non-profit centre staff are underpaid relative to

their qualifications and the situation is worsening with the decline in the real value of operating grants.

5 Morris (1990: 76) argues that the existence of such tax allowances has an important impact on the labour force participation rate of women with very young children, which is much lower in countries such as Britain where no such allowances are permitted.

6 The illogicalities of the ideological and institutional separation between daycare and schooling for four- and five-year-olds are becoming more evident: for as daycare centres become increasingly professionalized and pedagogical in orientation, the overlap in what these children do in daycare and kindergarten is increasing. Moreover, it has been proposed in Ontario that the public kindergarten network be extended both temporally and geographically to help alleviate the daycare crisis (*Globe and Mail* 1989).

7 See Sorrentino (1990: 54) for comparative data on state childcare expenditures as a percentage of gross domestic product in Canada, the USA, Japan and major West European countries.

8 The details of programme implementation were left up to the provinces; thus, for instance, financial aid eligibility scales varied considerably and some provinces (e.g. Ontario and Manitoba) tied aid to parents to attendance at a non-profit centre, while others (e.g. Québec) let parents use their direct subsidy at any licensed service. Prior to the federal legislation, interprovincial variations had been much greater. The province of Québec, which collects its own income taxes, also allows deductions for childcare costs.

9 In Québec, operating grants seem to make little difference to the cost of care to parents using non-profit centres, compared to commercial ones which are not subsidized, but, as mentioned earlier, the quality of care is likely to be higher in the former.

10 For-profit centres are gaining ground, however, and proposed policy changes may accelerate this trend (Centrale d'Enseignement du Québec 1989).

11 In a comment very pertinent to the Québec case, Prentice (1989) argues that the most important distinction between profit and non-profit centres is an ideological one: the 'real' non-profit centres are parent-controlled cooperatives which seek to redefine the concept of daycare by providing an alternative to its commodification and/or bureaucratization.

12 The pre-school figures are comparable to those for Canada as a whole (Mackenzie and Truelove 1991) and for the USA (Wash and Brand 1990). For school daycare services, Québec's levels of provision are higher than other provinces (Kuiken 1986).

13 The Québec government is, moreover, caught in the grips of its own version of what McDowell (1990) sees as a new contradiction between production and reproduction: the province has one of the world's lowest birth rates (about 1.4 per thousand), and increasing daycare provision is seen as one possible way of encouraging more women to have a second or third child – this is a major reason for advocating a 'family policy' (Ouellet 1989).

14 By means of a questionnaire distributed via thirteen schools, information was collected about childcare practices and parents' employment situations (including job type, schedules and place of employment, family structure, country of birth and immigration history). A response rate of 70 per cent was acheived. Follow-up interviews were carried out with groups of parents and key informants from ethnic minority communities (for details see Blanc, Chicoine and Germain 1989; Chicoine, Germain and Rose forthcoming;

Rose and Chicoine 1991). The present author administered the project and coordinated the research team. All team members were involved in research design and field work.

15 Several generations often live in a dwelling of three superimposed flats (triplex), each with separate entrance, owner-occupied by the family; even when these families move to the suburbs, grandparents also commonly accompany their adult children.

16 No research exists as to the locational strategies of commercial centres and the effects of this on accessibility to different groups. Casual observation in Montréal suggests that such centres typically locate on busy thoroughfares on transportation routes and/or offer private bus services.

17 Some hospitals have workplace daycare, but this type of arrangement is not appropriate for young school-aged children unless institutional arrangements are made to transport children between the workplace daycare and the school. The only example of this in Canada is in a mining community of 350 inhabitants in the north of Baffin Island (Mayfield 1990: 120–2) – where, interestingly, the pre-school daycare is located in the same complex as the school and most of the government services.

18 A rare exception is the recently set up Canadian Auto Workers Union childcare network in Windsor, Ontario, which has 24-hour care for children of different ages, offered within a coordinated network of formal and licensed informal services, available both at the workplace and near parents' homes, and including transportation arrangements (Mayfield 1990).

19 For instance, there is a tax credit which can be used instead of the childcare deduction, but it takes a good accountant to work out which is more advantageous in one's particular situation.

20 Calculated from data in Statistics Canada, *Census of 1986*, cat. 93–155, Table 6, Ottawa: Supply and Services Canada. Their job ghettoization is compounded by immigration regulations which usually deem them to be 'dependants' of primary wage earners and thus ineligible for childcare or training allowances that would enable them to take language classes (Seward and McDade 1988).

21 It is dangerous because New Right elements in government are pitting middle-class women in two-earner families against single mothers by asserting that 'yuppie couples' are using up scarce financial aid resources for daycare. Here, it is useful to keep a comparative context in mind: while the neoliberal childcare agenda will only help those defined as 'special needs groups' (see Clapham and Smith 1990), countries such as Sweden and Denmark believe in assisting *all* parents to find affordable quality childcare through mainstream rather than targeted policies (see Sorrentino 1990).

11 WOMEN'S TRAVEL PATTERNS AT VARIOUS STAGES OF THEIR LIVES

1 See the discussion in Sandra Rosenbloom, 'The growth of non-traditional families: a challenge to traditional planning approaches', in G. Jensen *et al.* (eds), *Transportation and Mobility in an Era of Transition*, Amsterdam: Elsevier, North Holland, 1985, pp. 77–80.

2 See, for example, William G. Bell, 'Introduction to special issue on mobility and transportation of single parents', *Journal of Specialized Transportation Planning and Practice*, Vol. 3, no. 3, 1989, pp. ii–iii; and Lidia P. Kostyniuk *et al.*, 'Mobility of single parents; what do the trip records show?' in

Journal of Specialized Transportation Planning and Practice, Vol. 3, no. 3, 1989, p. 207.

3 These studies were funded through grants made to the author by the German Marshall Fund of the United States, the Dutch Ministry of Transport, the French Centre National de Recherche Scientifique and the US National Science Foundation.

4 The NPTS data were organized from the tapes by Malcolm Quint and Rich Mariotta, then of the staff of the Transportation Research Board, in support of the Study on Improving Mobility and Safety for Older Persons, conducted by the National Research Council of the National Academy of Sciences. The author is grateful for the use of these data.

5 If the few non-car owning or driving respondents are removed from the sample, the percentages shown in Figures 11.1 and 11.2 do not change appreciably.

6 Car ownership was, again, almost universal for all respondents here.

7 See US Census, *Statistical Abstract*, 1988. Table 73 gives the 1986 figure for women living alone as 31.3 per cent, while Table 63 gives it as 41.3 per cent.

8 The percentage of all trips by car by women without licences actually goes up with age, rising from 40 per cent of those 66–70 to 71 per cent of those over 85 – the average for all women 65+ without licences is 56 per cent.

9 Data specifically on elderly women drivers living alone were not available; the analyses given rely on data mixing those living alone and with families.

10 See Mourant, R. R. and Rackoff, N. J. 'Driving performance of the elderly', *Accident Analysis and Prevention*, Vol. 11, 1979, pp. 243–53 and Risser, R. and Chaloupka, C. 'Elderly drivers: risks and their causes', in J. A. Rolhengatter and R. A. de Bruin (eds), *Proceedings of the Second International Conference on Road Safety*, Assen, The Netherlands, 1988.

11 There is some evidence that those without cars may travel to the grocery store by bus but take a taxi home from the shop.

12 Note that women at all income levels are currently more likely to use public transport or to join carpools than men with comparable personal or household income.

12 WOMEN, THE STATE AND THE LIFE COURSE IN URBAN AUSTRALIA

The support of the Australian Research Council for this research is gratefully acknowledged.

1 Fincher and McQuillen (1989), for example, review the participation of women in urban social movements, including some commentary on women's experience with state bureaucracies.

2 This concern emerges in particular because the federal government is cutting the funds it provides to states and curbing the borrowing that might have allowed states to keep up the maintenance and extension of infrastructure (see Murphy *et al.* 1990; Burke and Hayward 1990).

REFERENCES

1 WHEN IN THE WORLD ARE WOMEN?

Aitken, S. C., Cutter, S., Foote, K. E. and Sell, J. L. (1989) 'Environmental perception and behavioral geography', in G. L. Gaile and C. Willmott (eds) *Geography in America*, Columbus, OH: Merrill Publishing Company.

Allatt, P., Keil, T., Bryman, A. and Bytheway, B. (eds) (1987) *Women and the Life Cycle: Transitions and Turning Points*, New York: St Martin's Press.

Anstee, M. J. (1990) 'Population aging in Latin America and the Caribbean: implications for policy makers and practitioners', in I. Hoskins (ed.) *Coping with Social Change: Programs that Work*, Washington, D.C.: American Association of Retired Persons in Association with the International Federation on Ageing.

Back, K. W. (1980) 'Introduction', in K.W. Back (ed.) *Life Course: Integrative Theories and Exemplary Populations*, Boulder, Colorado: Westview Press for the American Association for the Advancement of Science.

Baruch, G. and Brooks-Gunn, J. (1984) *Women in Midlife*, New York: Plenum.

Bateson, M.C. (1990) *Composing a Life*, New York: Plume/Penguin Books.

Beneria, L. (ed.) (1982) *Women and Development: the Sexual Division of Labor in Rural Societies*, New York: Praeger.

Berger, J. (1974) *The Look of Things*, New York: Viking.

Campbell, R. T., Abolafia, J. and Maddox, G. L. (1985) 'Life course analysis: use of replicated social surveys to study cohort differences', in A. Rossi (ed.) *Gender and the Life Course*, New York: Aldine Publishing.

Chaney, E. (1990) 'Empowerment of older women: evidence from historical and contemporary sources', in I. Hoskins (ed.) *Coping with Social Change: Programs that Work*, Washington, D.C.: American Association of Retired Persons in Association with the International Federation on Ageing.

Dinnerstein, M. (1992) *Between Two Worlds: Midlife Reflections on Work and Family*, Philadelphia: Temple University Press.

Featherman, D. L. (1983) 'Life-span perspectives in social science research', in P. B. Baltes and O. G. Brim, Jr. (eds) *Life-span Development and Behavior*, Vol. 5, New York: Academic Press.

Golant, S. M., Rowles, G. D. and Meyer, J. (1989) 'Aging and the aged', in G. L. Gaile and C. Willmott (eds) *Geography in America*, Columbus, OH: Merrill Publishing Company.

Hagerstrand, T. (1975) 'Survival and arena: on the life history of individuals in relation to their geographical environment', *The Monadnock* 49 (June): 9–29.

Hagerstrand, T. (1982) 'Diorama, path and project', *Tijdschrift voor Economische en Sociale Geografie* 73: 323–39.

Hareven, T. (1982) 'The life course and aging in historical perspective', in T. Hareven and K. J. Adams (eds) *Aging and Life Course Transitions: An Interdisciplinary Perspective*, New York: Guilford Press.

Hareven, T. and Adams K. J. (1982) 'Preface' in T. Hareven and K. J. Adams (eds) *Aging and Life Course Transitions: An Interdisciplinary Perspective*, New York: Guilford Press.

Hart, R. A. (1983) 'Children's geographies and the geography of children', in T. Saarinen, J. Sell and D. Seamon (eds) *Behavioral Geography: Prospect and Inventory*, Chicago: University of Chicago Monograph.

Johansson, S. and Nygren, D. (1991) 'The missing girls of China: a new demographic account', *Population and Development Review* 17, 1: 35–51.

Kristof, N. D. (1991) 'Stark data on women: 100 million are missing', *New York Times*, 5 November.

Lopata, H. Z. (1987) 'Women's family roles in life course perspective', in B. Hess and M. M. Ferre (eds) *Analyzing Gender: A Handbook of Social Science Research*, California: Sage.

Martensson, S. (1977) 'Childhood interaction and temporal organization', *Economic Geography* 53: 99–125.

Mesa-Lago, C. (1990) 'The current situation, limitations, and potential role of social security schemes for income maintenance and health care in Latin America and the Caribbean: focus on women', in I. Hoskins (ed.) *Coping with Social Change: Programs that Work*, Washington, D.C.: American Association of Retired Persons in Association with the International Federation on Ageing.

Momsen, J. H. and Townsend, J. (eds) (1987) *Geography of Gender in the Third World*, London: Hutchinson.

Monk, J. (1984) 'Human diversity and perceptions of place', in H. Haubrich (ed.) *Perceptions of People and Places through Media*, Freiburg: Commission on Geographical Education, International Geographical Congress.

Network News (1990) *Network News: A Newsletter of the Global Link for Midlife and Older Women* 5, 1: 9–15.

Neugarten B. (1985) 'Interpretive social science and research on aging', in A. Rossi (ed.) *Gender and the Life Course*, New York: Aldine.

Ozawa, M. N. (ed.) (1989) *Women's Life Cycle and Economic Insecurity*, New York: Greenwood Press.

Population Reference Bureau (1989) *Teen Mothers: Global Patterns*, Washington, D.C.: Population Reference Bureau.

Pred, A. (1981) 'Social reproduction and the time-geography of everyday life', *Geografiska Annaler* 63B: 1–22.

Rosser, S. V. (1991) 'Eco-feminism: lessons for feminism from ecology', *Women's Studies International Forum* 14, 3: 143–51.

Rossi, A. (1980) 'Life span theories and women's lives', *Signs* 6, 1: 4–32.

Rowles, G. D. (1981) 'Geographic perspectives on human development', *Human Development* 24: 67–76.

Saarinen, T., Sell, J. L. and Husband, E. (1982) 'Environmental perception: international efforts', *Progress in Human Geography* 6: 515–46.

Saraceno, C. (1991) 'Changes in life course patterns and behavior of three cohorts of Italian women', *Signs* 16, 3: 502–21.

Seager J. and Olson, A. (1986) *Women in the World: An International Atlas*, New York: Simon and Schuster.

Sivard, R. L. (1985) *Women . . . a World Survey*, Washington, D.C.: World Priorities.

Soja, E. W. (1989) *Postmodern Geographies*, London: Verso.

Stapleton, C. (1980) 'Reformulations of the family life-cycle concept: implications for residential mobility', *Environment and Planning A* 12: 1103–18.

UNICEF (1990) *State of the World's Children*, Oxford: Oxford University Press.

White, S., Brown, L. A., Clark, W. A. V., Gober, P., Jones, R., McHugh, K. and Morrill, R. (1989) 'Population geography', in G. L. Gaile and C. Willmott (eds) *Geography in America*, Columbus, OH: Merrill Publishing Company.

Zelinsky, W., Monk, J. and Hanson, S. (1982) 'Women and geography: a review and prospectus', *Progress in Human Geography* 6: 317–66.

2 WOMEN AND WORK ACROSS THE LIFE COURSE

Barrett, M. and McIntosh, M. (1982) *The Antisocial Family*, London: Verso.

Beechey, V. (1977) 'Some notes on female wage labor in capitalist production', *Capital and Class* 3: 45–66.

Bird, E. and West, J. (1987) 'Interrupted lives: a study of women returners', in P. Allatt, T. Keil, A. Bryman and B. Bytheway (eds) *Women and the Life Cycle: Transitions and Turning Points*, London: Macmillan.

California Federal Savings v. *Guerra*. 758 F.#2d 390 (C.A.#9 1985) affirmed 479 US 272.

Cunnison, S. (1987) 'Women's three working lives and trade-union participation', in P. Allatt, T. Keil, A. Bryman and B. Bytheway (eds) *Women and the Life Cycle: Transitions and Turning Points*, London: Macmillan.

Davis, K. (1988) 'Wives and work: a theory of the sex-role revolution and its consequences', in S. M. Dornbusch and M. H. Stroker (eds) *Feminism, Children, and the New Families*, New York: Guilford Press.

Elshtain, J. B. (1987) 'Feminist political rhetoric and women's studies', in J. S. Nelson, A. Megill and D. N. McCloskey (eds) *The Rhetoric of the Human Sciences: Language and Argument in Scholarship and Public Affairs*, Madison, Wisconsin: University of Wisconsin Press.

French, D. (1988) '60% of N.E. women work', *Boston Globe* 23 May: 22.

Fuchs, V. (1989) 'Mommy track is good for both business and families', *Wall Street Journal*, 13 March: A14.

Gage, N. (1989) *A Place For Us: Eleni's Children in America*, Boston: Houghton Mifflin.

Hanson, S. and Pratt, G. (1988) 'Reconceptualizing the links between home and work in urban geography', *Economic Geography* 64: 299–321.

Hanson, S. and Pratt, G. (1990) 'Dynamic dependencies: local employers, local lives in a global economy', paper given at Annual Meeting of the Regional Science Association, Boston, November.

Hanson, S. and Pratt, G. (1991) 'Job search and the occupational segregation of women', *Annals of the Association of American Geographers* 81: 229–53.

Heilbrun, C. G. (1988) *Writing a Woman's Life*, New York: Ballantine.

Lamphere, L. (1987) *From Working Daughters to Working Mothers: Immigrant Women in a New England Industrial Community*, Ithaca, New York: Cornell University Press.

McLaughlin, S. D., Melber, B. D., Billey, J. O. G., Zimmerle, D. M., Winges, L. D. and Johnson, T. R. (1988) *The Changing Lives of American Women*, Chapel Hill: University of North Carolina Press.

Nelson, K. (1986) 'Female labor supply characteristics and the suburbanization of low-wage office work', in M. Storper and A. Scott (eds) *Production, Work, Territory*, Boston: Allen and Unwin.

Parr, J. (1990) *The Gender of Breadwinners: Women, Men and Change in Two Industrial Towns, 1880 – 1950*, Toronto: University of Toronto Press.

Peterson, B. R. (1989) 'Firm size, occupational segregation, and the effect of family status on women's wages', *Social Forces* 68: 397–414.

Phillips, A. (1987) *Divided Loyalties: Dilemmas of Sex and Class*, London: Virago.

Pratt, G. and Hanson, S. (1991a) 'Time, space, and occupational segregation of women: a critique of human capital theory', *Geoforum* 22: 149–57.

Pratt, G. and Hanson, S. (1991b) 'On the links between home and work: family-household strategies in a buoyant labour market', *International Journal of Urban and Regional Research* 15: 55–74.

Pringle, R. (1989) *Secretaries Talk: Sexuality, Power and Work*, Toronto: Verso.

Redclift, N. (1985) 'The contested domain: gender, accumulation and the labour process', in N. Redclift and E. Mingione (eds) *Beyond Employment: Household, Gender and Subsistence*, Oxford: Basil Blackwell.

Reskin, B. (1991) 'Bringing the men back in: sex differentiation and the devaluation of women's work', in J. Lorbes and S. Farrell (eds) *The Social Construction of Gender*, Newbury Park, Ca.: Sage.

Rosenzweig, R. (1983) *Eight Hours For What We Will: Workers and Leisure in an Industrial City, 1870–1920*, Cambridge: Cambridge University Press.

Siltanen, J. (1986) 'Domestic responsibilities and the structuring of employment', in R. Crompton and M. Mann (eds) *Gender and Stratification*, Cambridge: Polity Press.

Smith, J. (1983) 'Feminist analyses of gender: a mystique', in M. Lowe and R. Hubbard (eds) *Women's Nature: Rationalization of Inequality*, New York: Pergamon.

Stacey, J. (1990) *Brave New Families: Stories of Domestic Upheaval in the Late Twentieth Century America*, New York: Basic Books.

Tilley, L. A. (1985) 'Family, gender, and occupation in industrial France: past and present', in A. S. Rossi (ed.) *Gender and the Life Course*, New York: Aldine.

Treiman, D. J. (1985) 'The work histories of women and men: what we know and what we need to find out', in A. S. Rossi (ed.) *Gender and the Life Course*, New York: Aldine.

Warshaw, M. (1991) 'Little Saigon, Central Mass', *Inside Worcester* May: 12–19.

3 ELIMINATING THE JOURNEY TO WORK

Christensen, K. (1986) 'Pros and cons of clerical homework', Testimony before the Employment and Housing Subcommittee, Committee on Government Operations, U.S. House of Representatives, Washington, D.C.

Christensen, K. (1988) *Women and Home-based Work: The Unspoken Contract*, New York: Henry Holt and Co.

Christensen, K. (1989) *Flexible Staffing and Scheduling in U.S. Corporations*, Research Bulletin No. 240, New York: The Conference Board, Inc.

Cichocki, M. (1980) 'Women's travel patterns in a suburban development', in G. R. Wekerle, R. Patterson and D. Morley (eds) *New Space for Women*, Boulder, CO: Westview Press.

Coutras, J., and Fagnani, J. (1978) 'Femmes et transports en milieu urbain', *International Journal of Urban and Regional Research* 2: 432–9.

Ericksen, J. (1977) 'An analysis of the journey to work for women', *Social Problems* 24: 428–35.

Friedan, B. (1963) *The Feminine Mystique*, New York: Dell.

Hirshey, G. (1985) 'How women feel about working at home', *Family Circle* 98, 15: 70–4.

Hochschild, A. with A. Machung (1989) *The Second Shift*, New York: Avon Books.

Horvath, F. (1986) 'Work at home: new findings from the current population survey', *Monthly Labor Review* 109: 31–5.

Kaniss, P. and Robins, B. (1974) 'The transportation needs of women', in K. Hapgood and J. Getzels (eds) *Women, Planning and Change*, Chicago: American Society of Planning Officials.

Madden, J. F. and White, J. W. (1978) 'Women's work trips: an empirical and theoretical overview', paper given at Women's Travel Issues Conference, sponsored by the U.S. Dept. of Transportation, Washington, D.C., September.

New York Times (1991) 'Census study finds drop in home ownership', June 16: 18.

Palm, R. and Pred, A. (1974) 'A time geographic perspective on problems of inequality for women', Working Paper No. 236, Institute of Urban and Regional Development, Berkeley.

4 GROWING GIRLS/CLOSING CIRCLES

Anderson, J. and Tindal, M. (1972) 'The concept of home range: new data for the study of territorial behavior', in W. Mitchell (ed.) *Environmental Design: Research and Practice*, Los Angeles: University of California Press.

Bartlett, S. (1991) '"Kids aren't like they used to be": nostalgia and reality in neighborhood life', *Children's Environments Quarterly* 8, 1: 49–58.

Bjorklid, P. (1985) 'Children's outdoor environment from the perspectives of environmental and developmental psychology', in T. Garling and J. Vaalsiner (eds) *Children within Environments: Toward a Psychology of Accident Prevention*, New York: Plenum.

Blakeley, K. (1987) 'Parents' conception of children's safety in neighborhood play settings', doctoral dissertation proposal, New York: City University of New York Subprogram in Environmental Psychology.

Bronfenbrenner, U. (1979) *The Ecology of Human Development*, Cambridge: Harvard University Press.

Bruner, J. S. and Connolly, K. J. (eds) (1974) *The Growth of Competence*, London: Academic Press.

Cain, M. T. (1977) 'The economic activities of children in a village in Bangladesh', *Population and Development Review* 3, 3: 301–27.

Carbonara-Moscati, V. (1985) 'Barriers to play activities in the city environment: a study of children's perceptions', in T. Garling and J. Vaalsiner (eds) *Children within Environments: Toward a Psychology of Accident Prevention*, New York: Plenum.

Douglas, M. and Wildavsky, A. (1982) *Risk and Culture*, Berkeley: University of California Press.

Edelman, M. W. (1991) Commencement address, Clark University, Worcester, MA. as published in *Clark News* 14, 2:1, 6.

Erikson, E. H. (1963) *Childhood and Society*, 2nd edn, New York: Norton.

Fernea, E. W. and Bezirgan, B. Q. (1977) *Middle Eastern Muslim Women Speak*, Austin: University of Texas Press.

Fischhoff, B., Lichtenstein, S., Slovic, P., Derby, S. and Keeney, R. L. (1981) *Acceptable Risk*, Cambridge: Cambridge University Press.

Galal el Din, M. (1977) 'The economic value of children in rural Sudan', in

REFERENCES

J. C. Caldwell (ed.) *The Persistence of High Fertility in the Third World*, Canberra: Australian National University, Department of Demography.

Harris, L. J. (1981) 'Sex-related variations in spatial skill', in L. S. Liben, A. H. Patterson and N. Newcombe (eds) *Spatial Representation and Behavior across the Life Span*, New York: Academic Press.

Hart, R. (1979) *Children's Experience of Place*, New York: Irvington.

Hart, R. (1986) 'The changing city of childhood: implications for play and learning', Catherine Molony Memorial Lecture, New York: City College Workshop Center.

Kates, R. W. (1977) 'Assessing the assessors: the art and ideology of risk assessment', *Ambio* 6, 5: 247–52.

Katz, C. (1986) 'Children and the environment: work, play and learning in rural Sudan', *Children's Environments Quarterly* 3, 4: 43–51.

Katz, C. (1989) 'Herders, gatherers and foragers: the emerging botanies of children in rural Sudan', *Children's Environments Quarterly* 6, 1: 46–53.

Katz, C. (1991a) 'Sow what you know: the struggle for social reproduction in rural Sudan', *Annals of the Association of American Geographers* 81, 3: 488–514.

Katz, C. (1991b) 'Cable to cross a curse: the everyday practices of resistance and reproduction among youth in New York City', unpublished MS.

Kirby, A. M. (ed.) (1990) *Nothing to Fear: Risks and Hazards in American Society*, Tucson: University of Arizona Press.

Lazreg, M. (1988) 'Feminism and difference: the perils of writing as a woman on women in Algeria', *Feminist Studies* 14, 1: 81–107.

Liben, L. S. (1981) 'Spatial representation and behavior: multiple perspectives', in L. S. Liben, A. H. Patterson and N. Newcombe (eds) *Spatial Representation and Behavior across the Life Span*, New York: Academic Press.

Longhurst, R. (1982) 'Resource allocation and the sexual division of labour: a case study of a Moslem Hausa village in Northern Nigeria', in L. Beneria (ed.) *Women and Development: The Sexual Division of Labor in Rural Societies*, New York: Praeger.

Mascarenhas, A. (1977) *The Participation of Children in Socio-Economic Activities: The Case of Rukwa Region*, Dar es Salaam: University of Dar es Salaam, BRALUP. Research Report 20–1.

Matthews, M. (1987) 'Gender, home range and environmental cognition', *Transactions of the Institute of British Geographers* 12, 1: 43–56.

Mies, M. (1982) 'The dynamics of the sexual division of labour and integration of rural women into the world market', in L. Beneria (ed.) *Women and Development: The Sexual Division of Labor in Rural Societies*, New York: Praeger.

Monk, J. (1992) 'Gender in the landscape: expressions of power and meaning', in K. Anderson and F. Gale (eds) *Inventing Places: Studies in Cultural Geography*, Melbourne: Longmans Cheshire.

Munroe, R. H. and Munroe, R. L. (1971) 'Effect of environmental experience on spatial ability in an East African society', *Journal of Social Psychology* 83: 15–22.

Nag, M., White, B. N. F. and Peet, R. C. (1978) 'An anthropological approach to the study of the economic value of children in Java and Nepal', *Current Anthropology* 19, 2: 293–306.

Nerlove, S. B., Munroe, R. H. and Munroe, R. L. (1971) 'Effect of environmental experience on spatial ability: a replication', *Journal of Social Psychology* 84: 3–10.

Nerlove, S. B., Roberts, J. M., Klein, R. E., Yarbrough, C. and Habicht, J.-P. (1974) 'Natural indicators of cognitive development: an observational study of rural Guatemalan children', *Ethos*.

Pain, R. (1991) 'Space, sexual violence and social control: integrating geographical and feminist analyses of women's fear of crime', *Progress in Human Geography* 15, 4: 415–31.

Robinson, B. E., Coleman, M. and Rowland, B. H. (1986) 'The after-school ecologies of latchkey children', *Children's Environments Quarterly* 3, 2: 4–8.

Rodgers, G. and Standing, G. (1981) *Child Work, Poverty and Underdevelopment*, Geneva: International Labour Organisation.

Ruddle, K. and Chesterfield, R. (1977) *Children's Learning for Food Procurement on the Orinoco Delta*, Los Angeles and Berkeley: University of California Press.

Saegert, S. and Hart, R. (1978) 'The development of environmental competence in girls and boys', in M. Salter (ed.) *Play: Anthropological Perspectives*, Cornwall, NY: Leisure Press.

Schildkrout, E. (1981) 'The employment of children in Kano (Nigeria)', in G. Rodgers and G. Standing (eds), *Child Work, Poverty and Underdevelopment*, Geneva: International Labour Organisation.

Self, C. M., Gopal, S., Golledge, R. G. and Fenstermaker, S. (1991) 'Gender-related differences in spatial abilities', unpublished manuscript, University of California, Santa Barbara.

Smith, P. (1991) Children's Defense Fund (personal communication).

Swedenburg, T. (1989) 'Palestinian women in the 1936–39 revolt: implications for the Intifada', paper given at 'Marxism Now: Tradition and Difference' conference, University of Massachusetts, Amherst, November.

Tindal, M.A. (1971) *The Home Range of Black Elementary School Children: An Exploratory Study in the Measurement and Comparison of Home Range*, Worcester, MA: Graduate School of Geography, Clark University. Place Perception Research Report 8.

Torell, G. and Biel, A. (1985) 'Parental restrictions and children's acquisition of neighborhood knowledge', in T. Garling and J. Vaalsiner (eds) *Children within Environments: Toward a Psychology of Accident Prevention*, New York: Plenum.

Valentine, G. (1989) 'The geography of women's fear', *Area* 21, 4: 385–90.

White, R. (1990) *No Space of their Own: Young People and Social Control in Australia*, Cambridge: Cambridge University Press.

5 'HE WON'T LET SHE STRETCH SHE FOOT'

Armstrong, D. (1990) *The Old Village and the Greathouse: An Archaeological and Historical Examination of Drax Hall Plantation, St. Ann's Bay, Jamaica*, Urbana and Chicago: University of Illinois Press.

Austin, D. (1974) 'Symbols and ideologies of class in urban Jamaica: a cultural analysis of classes', unpublished Ph.D. dissertation, University of Chicago.

Bridenbaugh, C. and Bridenbaugh, R. (1972) *No Peace Beyond the Line: The English in the Caribbean 1624–1690*, New York: Oxford University Press.

Brodber, E. (1975) *A Study of Yards in the City of Kingston*, Mona, Jamaica: Institute of Social and Economic Research, Working Papers, no. 9.

Clarke, E. (1966) *My Mother Who Fathered Me: A Study of the Family in Three Selected Communities in Jamaica*, London: George Allen & Unwin.

Craton, M. (1978) *Searching for the Invisible Man: Slaves and Plantation Life in Jamaica*, Cambridge: Harvard University Press.

Dunn, R. (1972) *Sugar and Slaves*, Chapel Hill: University of North Carolina Press.

REFERENCES

Edwards, B. (1793) *The History, Civil and Commercial, of the British Colonies in the West Indies,* 2 vols, London: John Stockdale.

Greenfield, S. (1966) *English Rustics in Black Skins,* New Haven: College and University Publishers.

Handler J. S. and Lange F. W. (1978) *Plantation Slavery in Barbados: An Archaeological and Historical Investigation,* Cambridge MA: Harvard University Press.

Higman, B. (1974) 'A report on the excavations at Montpelier and Roehampton', *Jamaica Journal* 8.

Higman, B. (1975) 'The slave family and household in the British West Indies', *Journal of Interdisciplinary History* 6: 261–87.

Howson, J. (1987) 'Report of the field study of Galways Village', unpublished report of the Galways Plantation Project, Montserrat, West Indies.

Jagdeo, T. P. (1984) *Teenage Pregnancy in the Caribbean,* New York: International Planned Parenthood Federation.

Jopling, C. F. (1988) *Puerto Rican Houses in Sociohistorical Perspective,* Knoxville: University of Tennessee Press.

Olwig, K.F. (1985) *Cultural Adaptation and Resistance on St. John: Three Centuries of Afro-Caribbean Life,* Gainesville: University of Florida Press.

Patterson, O. (1967) *The Sociology of Slavery,* London and Rutherford: Fairleigh Dickenson University Press.

Pulsipher, L. M. (1986) 'Ethnoarchaeology for the study of Caribbean slave villages', Symposium on Ethnohistory and Historical Archaeology, Baltimore: Johns Hopkins University.

Pulsipher, L. M. (1992) 'Changing roles in the life cycles of women in traditional West Indian houseyards', in J. H. Momsen (ed.) *Women and Change: A Pan-Caribbean Perspective,* London: James Currey.

Pulsipher, L. M. and Wells-Bowie, L. (1989) 'The domestic spaces of Daufuskie and Montserrat: a cross-cultural comparison', *Traditional Dwellings and Settlements Series* No. 7, Berkeley: University of California.

Roberts, L. J. and Stefani, R. L. (1949) *Patterns of Living in Puerto Rican Families,* Rio Piedras, Puerto Rico: The University of Puerto Rico.

Smith, M. G. (1962) *West Indian Family Structure,* Seattle: University of Washington Press.

Smith, R. T. (1963) 'Culture and social structure in the Caribbean', *Comparative Studies in Society and History* 6: 24–46.

Smith, R. T. (1973) 'The matrifocal family', in J. Goody (ed.) *The Character of Kinship,* Cambridge: Cambridge University Press.

Smith, R. T. (1987) 'Hierarchy and the dual marriage system in West Indian society', in J. Collier and S. Yanagisako (eds) *Gender and Kinship: Essays Toward Unified Analysis,* Stanford: Stanford University Press.

Smith, R. T. (1988) *Kinship and Class in the West Indies,* Cambridge: Cambridge University Press.

Stewart, J. M. (1823) *A View of the Past and Present State of the Island of Jamaica,* Edinburgh: Oliver and Boyd.

Turner, V. W. (1957) *Schism and Continuity in an African Society: A Study of Ndembu Village Life,* Manchester: Manchester University Press.

Wentworth, T. (1834) *West India Sketch Book,* 2 vols, London: Whittaker and Company.

Westmacott, R. (1990) *Traditional Gardens and Yards of African-Americans in the Rural South,* Athens GA: University of Georgia School of Environmental Design.

Young, W. (1793) 'A tour through the several islands of Barbados, St. Vincent, Antigua and Grenada', in B. Edwards (ed.) *The History, Civil and Commercial, of the British Colonies in the West Indies*, London: John Stockdale.

6 WOMEN, WORK AND THE LIFE COURSE IN THE RURAL CARIBBEAN

Alcraft, R. (1987) 'A case study of beach vendors in Barbados: their role and potential in society and tourism', unpublished undergraduate dissertation, Department of Geography, University of Newcastle upon Tyne.

Berleant-Schiller, R. and Maurer, W. M. (1992) 'Women's place is every place: merging domains and women's roles in Barbuda and Dominica', in J. H. Momsen (ed.) *Women and Change: A Pan-Caribbean Perspective*, London: James Currey.

Besson, J. (1987) 'Family land in the Caribbean', in J. Besson and J. Momsen (eds) *Land and Development in the Caribbean*, London: Macmillan.

Brana-Shute, R. (1992) 'Neighbourhood networks and national politics among working class Afro-Surinamese women', in J. H. Momsen (ed.) *Women and Change: A Pan-Caribbean Perspective*, London: James Currey.

Caricom (n.d.a) *1980–1981 Population Census of the Commonwealth Caribbean. Montserrat, Vol. 1*, Jamaica: Statistical Institute.

Caricom (n.d.b) *1980–1981 Population Census of the Commonwealth Caribbean. Barbados, Vol. 3*, Jamaica: Statistical Institute.

Henshall (Momsen), J. (1981) 'Women and small scale farming in the Caribbean', in O. Horst (ed.) *Papers in Latin American Geography in Honor of Lucia C. Harrison*, Muncie, Indiana: Conference of Latin Americanist Geographers.

Jean, R. F. (1986) 'Towards a recognition of the "invisible" women in agriculture and rural development in St. Lucia', unpublished M.A. paper, School of Rural Planning and Development, University of Guelph.

Kelly, D. (1987) *Hard Work, Hard Choices: A Survey of Women in St. Lucia's Export-Oriented Electronics Factories*, Occasional Paper No. 20, Institute of Social and Economic Research, University of the West Indies, Barbados.

Lewis, W. A. (1950) *The Industrialization of the West Indies*, Reprint, Barbados: Government Printing Office.

Massiah, J. (1984) *Employed Women in Barbados: A Demographic Profile, 1946–1970*, Occasional Paper No. 8, Institute of Social and Economic Studies, University of the West Indies, Barbados.

Momsen, J. (1986) 'Migration and rural development in the Caribbean', *Tijdschrift voor Economische en Sociale Geografie* 77, 1: 50–8.

Momsen, J. (1987) 'Land settlement as an imposed solution', in J. Besson and J. Momsen (eds) *Land and Development in the Caribbean*, London: Macmillan.

Momsen, J. (1988a) 'Changing gender roles in Caribbean peasant agriculture', in J. S. Brierley and H. Rubenstein (eds) *Small Farming and Peasant Resources in the Caribbean*, Winnipeg, Manitoba: University of Manitoba Department of Geography.

Momsen, J. (1988b) 'Gender and labour displacement in agriculture in England', paper given at the International Geographical Union Congress, Sydney, Australia, August.

Reddock, R. (1989) 'Historical and contemporary perspectives: the case of Trinidad and Tobago', in K. Hart (ed.) *Women and the Sexual Division of Labour in the Caribbean*, Kingston, Jamaica: Consortium Graduate School of Social Sciences.

Skelton, T. (1989) 'Gender relations in Montserrat', unpublished Ph.D. dissertation, Department of Geography, University of Newcastle upon Tyne.

Spence, E. J. (1964) 'The hawker and the huckster in the Scotland District, Barbados', unpublished M.A. thesis, McGill University, Montreal.

7 GENDER AND THE LIFE COURSE ON THE FRONTIERS OF SETTLEMENT IN COLOMBIA

Chambers, R. (1969) *Settlement Schemes in Tropical Africa*, London: Routledge and Kegan Paul.

Deere, C. D. and Leon de Leal, M. (1983) *Women in Andean Agriculture*, Geneva: ILO.

Hecht, S. B. and Cockburn, A. (1990) *The Fate of the Forest*, London: Penguin.

Kearney, R. N. and Miller, B. D. (1983) 'Sex differential patterns of internal migration in Sri Lanka', *Peasant Studies* 10, 4: 233–50.

Meertens, D. (1988) *La Mujer Campesina en la Colonizacion del Guaviare (Colombia)*, Bogota: Informe Tecnico presentado al Proyecto DAINCO/CASAM, Corporacion de Araracuara.

Mies, M. (1986) *Patriarchy and Accumulation on a World Scale*, London: Zed.

Momsen J. and Kinnaird V. (eds) (forthcoming) *Regional Perspectives on Women and Development*, London: Routledge.

Momsen J. and Townsend J. G. (eds) (1987) *Geography of Gender in the Third World*, London: Hutchinson Educational.

Moser, C. O. N. (1987) 'Women, rural settlements and housing: a conceptual framework for analysis and policy-making', in C. O. N. Moser and L. Peake (eds) *Women, Human Settlements and Housing*, London: Tavistock.

Townsend, J. G. (forthcoming a) 'Housewifization in the Colombian rainforest', in J. H. Momsen and V. Kinnaird (eds) *Different Places, Different Voices*, London: Routledge.

Townsend, J. G. (forthcoming b) 'Geography and gender in agricultural colonization', in *Documents d'Analisi Geografica*, Barcelona.

Townsend, J. G. and Wilson de Acosta, S. (1987) 'Gender roles in the colonization of rainforest: a Colombian case study', in J. Momsen and J. G. Townsend (eds) *Geography of Gender in the Third World*, London: Hutchinson Educational.

Young, K. (1982) 'The creation of a relative surplus population: a case study from Mexico', in L. Beneria (ed.) *Women and Development: The Sexual Division of Labor in Rural Societies*, New York: Praeger.

8 OLD TIES

Bailey, K. A. (1985) 'A judicious mixture: Negroes and immigrants in the West Virginia mines, 1880–1917', in W. H. Turner and E. J. Cabbell (eds) *Blacks in Appalachia*, Lexington: University of Kentucky Press.

Benston, M. (1969) 'A political economy of women's liberation', *Monthly Review* 21, 4.

Corbin, D. A. (1981) *Life, Work and Rebellion in the Coal Fields, The Southern West Virginia Miners 1880–1922*, Urbana: University of Illinois Press.

Dalla Costa, M. (1972) 'Women and the subversion of the community', *Radical America* 6, 1: 67–102.

Densmore, R. E. (1977) *The Coal Miner of Appalachia*, Parsons, West Virginia: McClain Printing Company.
Edholm, F., Harris, O. and Young, K. (1978) 'Conceptualizing women', *Critique of Anthropology* 3, 9–10: 101–30.
Eller, R. (1982) *Miners, Millhands and Mountaineers: The Industrialization of the Appalachian South, 1880–1930*, Knoxville: University of Tennessee Press.
Secombe, W. (1974) 'The housewife and her labor under capitalism', *New Left Review* 83: 3–24.

9 LIFE COURSE AND SPACE

Bessy, P. (1987) 'Paris, un puzzle social', *Regards sur l'Île-de-France* 1: 6–14.
Blumen, O. and Kellerman, A. (1990) 'Gender differences in commuting distance, residence and employment location: Metropolitan Haifa 1972 and 1983', *Professional Geographer* 42, 1: 54–71.
Bonvalet, C. and Fribourg, A. M. (1990) *Stratégies Résidentielles*, Paris: INED, Plan Construction et Architecture.
Castelain-Meunier, C. and Fagnani, J. (1988a) 'Deux ou trois enfants: les nouveaux arbitrages des femmes', *Revue Française des Affaires Sociales* 1: 45–66.
Castelain-Meunier, C. and Fagnani, J. (1988b) *Avoir Deux ou Trois Enfants: Contraintes, Arbitrages et Compromis*, Paris: Caisse Nationale des Allocations Familiales.
Chauviré, Y. (1988) 'Les particularités des ménages parisiens', *Espace, Populations, Sociétés* 1: 141–6.
Fagnani, J. (1986) 'La durée des trajets quotidiens: un enjeu pour les mères actives', *Economie et Statistique* 185: 47–55.
Fagnani, J. (1990) 'City size and mothers' labour force participation', *Tijdschrift voor Economische en Sociale Geografie* 81, 3: 182–8.
Faure, A. (1990) 'De l'urbain à l'urbain: du courant parisien de peuplement en banlieue, 1880–1914', *Villes en Parallèle* 15–16: 155–73.
Gotman, A. (1989) *Héritier*, Paris: Presses Universitaires de France.
Hanson, S. and Johnson, I. (1985) 'Gender differences in work-trip length: explanations and implications', *Urban Geography* 3: 193–219.
Hanson, S. and Pratt, G. (1988) 'Reconceptualizing the links between home and work in urban geography', *Economic Geography* 64, 4: 299–321.
Hanson, S. and Pratt, G. (1990) 'Geographic perspectives on the occupational segregation of women', *National Geographic Research* 6, 4: 376–99.
Hantrais, L. (1990) *Managing Professional and Family Life: A Comparative Study of British and French Women*, Brookfield, VT: Dartmouth.
Institut National de la Statistique et des Etudes Economiques (1984) *Housing Inquiry*, Paris: INSEE.
Institut National de la Statistique et des Etudes Economiques (1988) *Housing Inquiry*. Paris: INSEE.
Johnson-Anumonwo, I. (1988) 'The journey to work and occupational segregation', *Urban Geography* 9, 2: 138–54.
Le Jeannic, T. (1990) 'Se tourner le dos le matin ou partir ensemble', *Regards sur l'Île-de-France* 8: 6–9.
Madden, J. F. (1981) 'Why women work closer to home', *Urban Studies* 18: 181–94.
Massot, A. (1990) 'Qui peut acheter un logement à Paris?', *Etudes Fonciéres* 48: 12–14.
Michelson, W. (1977) *Environmental Choice, Human Behavior and Residential Satisfaction*, New York: Oxford University Press.
Michelson, W. (1988) 'Divergent convergence: the daily routines of employed

spouses as a public affair agenda', in C. Andrew and B. M. Milroy (eds) *Life Spaces*, Vancouver: University of British Columbia Press.

Neitzert, F. (1990) *La Chambre d'Enfant: Représentations et Pratiques*, Paris: Plan Construction.

O'Donnel, L. and Stueve, A. (1981) 'Employed women: mothers and good neighbors', *Urban and Social Change Review* 14: 21–6.

Pinçon, M., Pinçon-Charlot, M. and Rendu, P. (1986) *Ségrégation urbaine*, Paris: Anthropos.

Pleck, J. H. (1985) *Working Wives/Working Husbands*, Beverly Hills, CA: Sage.

Rossi, A. (ed.) (1985) *Gender and the Life Course*, New York: Aldine.

Saegert, S. (1981) 'Masculine cities and feminine suburbs: polarized ideas, contradictory reality', in C. R. Stimpson *et al.* (eds) *Women and the American City*, Chicago: University of Chicago Press.

Shlay, A. and Di Gregorio, D. A. (1985) 'Same city, different worlds', *Urban Affairs Quarterly* 21, 1: 66–86.

Stapleton, C. (1980) 'Reformulations of the family life-cycle concept: implications for residential mobility', *Environment and Planning A* 12: 1103–18.

Taffin, C. (1987) 'L'accession à tout prix', *Economie et Statistique* 202: 24–38.

Team 'Espace, Population, Société' (1989) *Atlas Démographique et Social de l'Île-de-France*, Paris: STRATES, Direction Régionale de l'Equipement de la Région Île-de-France.

Thompson, L. and Walker, A. J. (1989) 'Gender in families: women and men in marriage, work and parenthood', *Journal of Marriage and the Family* 51: 845–71.

Villeneuve, P. and Rose, D. (1988) 'Gender and the separation of employment from home in Metropolitan Montreal, 1971–1981', *Urban Geography* 9, 2: 155–79.

Weiss, R. S. (1987) 'Men and their wives' work', in F. J. Crosby (ed.) *Spouse, Parent, Worker: On Gender and Multiple Roles*, New Haven, CT: Yale University Press.

10 LOCAL CHILDCARE STRATEGIES IN MONTRÉAL, QUÉBEC

Adamson, N., Briskin, L. and McPhail, M. (1988) *Feminist Organizing for Change: The Contemporary Women's Movement in Canada*, Toronto: Oxford University Press.

Allen, S. (1979) 'Pre-school children: ethnic minorities in England', *New Community* 8: 135–42.

Association des services de garde en milieu scolaire du Québec (1989) *Mémoire sur l'Enoncé de Politique sur les Services de Garde à l'Enfance*, Longueuil, QC.

Banting, K. G. (1987) 'The welfare state and inequality in the 1980s', *Canadian Review of Sociology and Anthropology/Revue canadienne de sociologie et d'anthropologie* 24, 3: 309–38.

Bernèche, F. pour le Regroupement des Organismes de Montréal Ethnique pour le Logement (ROMEL) (1990) *Problématique du Logement des Ménages Formant la Nouvelle Immigration à Montréal*, Montréal: Ville de Montréal, Service de l'Habitation et du Développement Urbain.

Blanc, B., Chicoine, N. and Germain, A. (1989) 'Quartiers multiethniques et pratiques familiales: la garde des jeunes enfants d'âge scolaire', *Revue Internationale d'Action Communautaire* 21, 61: 165–76.

Borchorst, A. and Sim, B. (1987) 'Women and the advanced welfare state – a new kind of patriarchal power?', in A. S. Sassoon (ed.) *Women and the State*, London: Hutchinson.

Bourdieu, P. (1984) *Distinction: A Social Critique of the Judgement of Taste*, London: Routledge.

Bracken, D. C., Hudson, P. and Selinger, G. (1988) 'Day care in Manitoba and the 1988 provincial election: the potential for change', *Nouvelles Pratiques Sociales* 1, 1: 171–80.

Canada, National Council on the Status of Women (1986) *Report of the Task Force on Child Care* (K, Cooke, chairperson), cat. no. SW41-1-1986, Ottawa: Supply and Services Canada.

Canada, National Council on Welfare (1988) *Child Care: A Better Alternative*, cat. no. H68-20-1988E, Ottawa: Supply and Services Canada.

Canadian Dimension (1989) Editorial: 'Mulroney's family', 23, 2: 3.

Chicoine, N., Germain, A. and Rose, D. (forthcoming) 'From economic restructuring to the fabric of everyday life: families' use of childcare services in multiethnic neighbourhoods in transition', in F.W. Remiggi (ed.) *The Changing Urban Geography of Montréal*, Victoria, BC: University of Victoria, Western Geographical Series.

Christian, P. B. (1990) 'Determinants of child care arrangements: shift work as a type of childcare', paper given at Annual Meetings of the Population Association of America, Toronto.

Christopherson, S. (1989) 'Flexibility in the US service economy and the emerging spatial division of labour', *Transactions of the Institute of British Geographers*, New Series, 14, 2:131–43.

Clapham, D. and Smith, S. J. (1990) 'Housing policy and "special needs"', *Policy and Politics* 18, 3: 193–205.

Croft, S. (1986) 'Women, caring and the recasting of need – a feminist reappraisal', *Critical Social Policy* 6, 1: 23–39.

Dale, J. and Foster, P. (1986) *Feminists and State Welfare*, London: Routledge.

Esping-Anderson, G. (1989) 'The three political economies of the welfare state', *Canadian Review of Sociology and Anthropology/Revue canadienne de sociologie et d'anthropologie* 26, 1: 9–36.

Finch, J. (1986) 'Lessons from Norway: women in a welfare society', *Critical Social Policy* 6, 1: 129–33.

Finch, J. (1989) *Family Obligations and Social Change*, Cambridge: Polity.

Fincher, R. (1989) 'Class and gender relations in the local labour market and the local state', in J. Wolch and M. Dear (eds) *The Power of Geography: How Territory Shapes Social Life*, London and Boston: Unwin Hyman.

Franzway, S., Court, D. and Connell, R. W. (1989) *Staking a Claim: Feminism, Bureaucracy and the State*, Sydney: Allen & Unwin.

Friendly, M., Cleveland, C. and Willis, T. (1989) *Flexible Child Care in Canada*, Toronto: University of Toronto, Centre for Urban and Community Studies, Childcare Resource and Research Unit.

Godbout, J. and Collin, J.-P. (1977) *Les Organismes Populaires en Milieu Urbain: Contre-pouvoir ou Nouvelle Pratique Professionnelle?* Montréal: INRS-Urbanisation, Rapports de recherches, No. 3.

Granger, D. (1987) 'Réflexion sur les enjeux sociaux de la politique québécoise des garderies des années 1970 à 1982', *Sociologie et sociétés* 19, 1: 73–81.

Greater London Council (1986) *The London Industrial Strategy*, London.

Grubb, W. N. and Lazerson, M. (1982) *Broken Promises: How Americans Fail their Children*, New York: Basic Books.

REFERENCES

Hanson, S. and Pratt, G. (1988) 'Reconceptualizing the links between home and work in urban geography', *Economic Geography* 64, 4: 300–21.

Harrison, M. L. (1986) 'Consumption and urban theory: an alternative approach based on the social division of welfare', *International Journal of Urban and Regional Research* 10, 2: 232–42.

Hatchuel, G. (1989) 'Accueil de la petite enfance et activité féminine', in Caisse Nationale des Allocations Familiales, *Recherche CNAF 1989*, Paris.

International Labour Office (1988) *Conditions of Work Digest* 7, 2.

Joffe, C. (1983) 'Why the United States has no child-care policy', in I. Diamond (ed.) *Families, Politics, and Public Policy*, New York: Longman.

Johnson, N, (1989) 'The privatization of welfare', *Social Policy and Administration* 23, 1: 17–30.

Kuiken, J. (1986) 'Latchkey children', unpublished report prepared for the House of Commons Special Committee on Child Care, Ottawa.

Le Bourdais, C. and Rose, D. (1986) 'Les familles monoparentales et la pauvreté', *Revue internationale d'action communautaire* 16, 56: 181–8.

Le Grand, J. and Robinson, R. (1984) 'Privatisation and the welfare state: an introduction', in J. Le Grand and R. Robinson (eds) *Privatisation and the Welfare State*, London: Allen and Unwin.

Léger, M. (1986) *Les Garderics: le Fragile Equilibre du Pouvoir. Les Enjeux d'une Gestion Communautaire*, Montréal: Regroupement des garderies de Montréal métropolitain, inc. and Éditions de l'Arche.

Lero, D. and Kyle, I. (1989) *Families and Children in Ontario: Supporting the Parenting Role*, Toronto: Child, Youth and Family Policy Research Centre, Discussion Paper.

Lewis, J. (1984) *Women in England, 1870–1950*, Brighton, UK: Wheatsheaf.

Lightman, E. and Irving, A. (1991) 'Restructuring Canada's welfare state', *Journal of Social Policy* 20, 1: 65–86.

Mackenzie, S. and Truelove, M. (1991) 'Change in access to public and private services: the case of childcare', in L. S. Bourne and D. F. Ley (eds) *The Changing Social Geography of Canadian Cities*, Montréal: McGill-Queen's.

Mann, S. (1986) 'Family, class and state in women's access to abortion and day care: the case of the United States', in J. Dickinson and B. Russell (eds) *Family, Economy and State: Social Reproduction Processes Under Capitalism*, New York: St Martin's Press.

Mark-Lawson, J., Savage, M. and Warde, A. (1985) 'Gender and local politics: struggles over welfare policies, 1918–1939', in Lancaster Regionalism Group *Localities, Class and Gender*, London: Pion.

Maroney, H. (1988) 'Using Gramsci for women: feminism and the Quebec state, 1960–1980', *Resources for Feminist Research/Documentation sur la recherche féministe* 17, 3: 26–30.

Martin, J. (1984) 'Non-English-speaking women: production and social reproduction', in G. Bottomley and M. M. de Lepervanche (eds) *Ethnicity, Class and Gender in Australia*, Sydney: Allen & Unwin.

Mayfield, M. I. (1990) *Work-Related Child Care in Canada*, cat. no. L38-42/90E, Ottawa: Labour Canada Women's Bureau.

McDowell, L. (1990) 'Production and reproduction in a post-Fordist era – a new contradiction?', paper given at Annual Meeting of the Association of American Geographers, Toronto, April.

Meintel, D., Labelle, M., Turcotte, G. and Kempeneers, M. (1985) 'La nouvelle double journée de travail des femmes immigrantes au Québec', *Revue internationale d'action communautaire* 14, 5: 33–44.

Moore, M. (1989) 'Dual-earner families: the new norm', Statistics Canada, cat. 11-008E, *Canadian Social Trends* 12: 24–6.

Morris, L. (1990) *The Workings of the Household: A US–UK Comparison*, Cambridge: Polity.

Myles, J. (1988) 'Decline or impasse? The current state of the welfare state', *Studies in Political Economy* 26: 73–107.

O'Connor, J. S. (1989) 'Welfare expenditure and policy orientation in Canada in comparative perspective', *Canadian Review of Sociology and Anthropology/ Revue canadienne de sociologie et d'anthropologie* 26, 1: 127–50.

Ontario, Ministry of Community and Social Services (1988) *Transitions: Report of the Social Assistance Review Committee*, Toronto: Queen's Printer for Ontario.

Ouellet, A. (1989) 'Les orientations majeurs de la politique familiale au Québec', in Québec, Conseil des Affaires sociales with the Secrétariat à la famille and the Bureau de la Statistique du Québec *Dénatalité: des Solutions. Colloque Internationale sur les Politiques Familiales*, Québec, QC: Les Publications du Québec.

Pascall, G. (1986) *Social Policy: A Feminist Analysis*, London: Tavistock.

Pence, A. R. (1987) 'Day care: changes in the role of the family and early childhood education', in L. L. Stewin and S. J. H. McCann (eds) *Contemporary Educational Issues: The Canadian Mosaic*, Toronto: Copp Clark.

Petit-à-Petit (1988) 'Dossier: les services de garde en milieu scolaire' 6, 6: 7–14.

Phillips, A. and Moss, P. (1989) *Who Cares for Europe's Children? The Short Report of the European Childcare Network*, Luxembourg: Office for Official Publications of the European Communities.

Prentice, S. (1988) 'The "mainstreaming" of daycare', *Resources for Feminist Research/Documentation sur la recherche féministe* 17, 3: 59–63.

Prentice, S. (1989) 'Childcare, commodification and the politics of auspices', paper given at Annual Meeting of the Canadian Sociology and Anthropology Association, Québec City, June.

Presse, La (1990) 'L'intégration passe par l'emploi', 5 May: A8.

Québec, Comité de la consultation sur la politique familiale (1985) *Collective Support Demanded for Québec Families. Report on the Consultation held on Family Policy, Part One*, Québec, QC: Ministère du Conseil exécutif, Secrétariat à la politique familiale.

Québec, Comité de la consultation sur la politique familiale (1986) *Collective Support Recommended for Québec Parents. Report of the Committee for Consultation on Family Policy, Part Two*, Québec, QC: Ministère du Conseil exécutif secrétariat à la politique familiale.

Québec, Comité ministériel permanent du développement social (1984) *For Québec Families. A Working Paper on Family Policy*, Québec, QC: Ministère des Affaires sociales.

Québec, Conseil du statut de la femme (1989) *Mémoire Présenté à la Commission Parlementaire des Affaires Sociales sur l'énoncé de Politique sur les Services de Garde à l'enfance: Pour un Meilleur Equilibre*, Québec, QC.

Québec, Ministère de l'Éducation et ministère de l'Enseignement supérieur (1989) *Répertoire des Organismes et Etablissements d'enseignement*, Édition 1989, Québec: Les Publications du Québec.

Québec, Ministère du Conseil exécutif, Ministre déléguée à la Condition féminine (1988) *Policy Statement on Day Care Services: A Better Balance, Orientation Paper*, Québec, QC.

Québec, Ministère du Conseil exécutif, Ministre déléguée à la Condition féminine (1989) 'Politique des services de garde à l'enfance: document explicatif; Les services de garde au Québec', unpublished. Sommaire: décisions finales.

Rose, D, (1990) 'Collective consumption revisited: analyzing modes of provision and access to childcare services in Montréal, Canada', *Political Geography Quarterly* 9, 4: 353–80.

Rose, D. and Chicoine, N. (1991) 'Access to school daycare services: class, family, ethnicity and space in Montréal's old and new "inner city"', *Geoforum* 22: 149–57.

Ruggie, M. (1984) *The State and Working Women: A Comparative Study of Britain and Sweden*, Princeton, NJ: Princeton University Press.

Sassoon, A. S. (1987) 'Women's new social role: contradictions of the welfare state', in A. S. Sassoon (ed.) *Women and the State: The Shifting Boundaries of Public and Private*, London: Hutchinson.

Seward, S. B. and McDade, K. (1988) *Immigrant Women in Canada: A Policy Perspective*, Background Paper BP 1988-1E, Ottawa: Canadian Advisory Council on the Status of Women.

Sgritta, G. B. (1989) 'Toward a new paradigm: family in the welfare state crisis', in K. Boh (ed.) for European Coordination Centre for Research and Documentation in Social Sciences *Changing Patterns of European Family Life: A Comparative Analysis of 14 European Countries*, London: Routledge.

Smith, D. E. and Griffith, A. I. (1989) 'Coordinating the uncoordinated: mothering, schooling and the family wage', unpublished paper, Department of Sociology in Education, Ontario Institute for Studies in Education.

Sorrentino, C. (1990) 'The changing family in international perspective', *Monthly Labor Review* 113, 3: 41–58.

Statistics Canada (1989a) *The Labour Force* (Dec.) cat. 71-002, Ottawa: Supply and Services Canada.

Statistics Canada (1989b) *Labour Force Survey: Annual Averages, 1981–1988*, cat. no. 71-529, Ottawa: Supply and Services Canada.

Taylor, P. (1988) 'Book review, P. Beresford and S. Croft, *Whose Welfare? Private Care or Public Services*, Brighton, UK: Lewis Cohen Urban Studies Centre, 1986', *Critical Social Policy* 8, 2: 122–5.

Waërness, K. (1989) 'Caring', in K. Boh (ed.) for European Coordination Centre for Research and Documentation in Social Sciences *Changing Patterns of European Family Life: A Comparative Analysis of 14 European Countries*, London: Routledge.

Wallman, S. (1984) *Eight London Households*, London and New York: Tavistock.

Wash, D. P. and Brand, L. E. (1990) 'Child day care services: an industry at a crossroads', *Monthly Labor Review* 113, 12: 17–24.

Yeandle, S. (1984) *Women's Working Lives: Patterns and Strategies*, London and New York: Tavistock.

11 WOMEN'S TRAVEL PATTERNS AT VARIOUS STAGES OF THEIR LIVES

Beroldo, S. (1991) 'Ridematching system effectiveness: a coast to coast perspective', Preprint 91-0227, paper presented to the 70th Annual Meeting of the Transportation Research Board, Washington, D.C., January.

Brainn, P. A. (1980) *Safety and Mobility Issues in Licensing and Education of Older Drivers*, Prepared for the US Department of Transportation, National Highway Safety Administration, Washington, D.C.

REFERENCES

California Department of Transportation (1990) *A Directory of California Trip Reduction Ordinances*, 2nd edn, Sacramento: Division of Transportation Planning, Technical Assistance Branch.

Comsis Corp (1989) *Evaluation of Travel Demand Management Measures to Relieve Congestion*, Report for the Federal Highway Administration, Washington, D.C., October.

Davidson, D. (1991) 'The impact of suburban employee trip chaining on transportation demand management', Preprint 91-0118, paper presented to the 70th Annual Meeting of the Transportation Research Board, Washington, D.C., January.

Deakin, E. (forthcoming) 'Land use and transportation planning in response to congestion: a review and critique', *Transportation Research Record*.

Dowling, R., Felham, D. and Wycko, W. (1991) 'Factors affecting TDM program effectiveness at six San Francisco medical institutions', Preprint 91-0136, Paper presented to the 70th Annual Meeting of the Transportation Research Board, Washington, D.C., January.

Ferguson, E. (1990) 'Transportation demand management: planning, development, and implementation', *Journal of the American Planning Association* 56, 4: 442–56.

Flynn, C. P. and Glazer, L. J. (1989) 'Ten cities' strategies for transportation demand management', *Transportation Research Record* 1212: 11–23.

Frederick, S. J. and Kenyon, K. L. (1991) 'The difficulty with "easy ride": obstacles to voluntary ridesharing in the suburbs', Preprint 91-0290, paper presented to the 70th Annual Meeting of the Transportation Research Board, Washington, D.C., January.

Giuliano, G. (forthcoming) 'Transportation demand management and urban traffic congestion: promise or panacea?', *Journal of the American Planning Association*.

Hanson, S. and Hanson, P. (1980) 'The impact of women's employment on household travel patterns: a Swedish example', in S. Rosenbloom (ed.) *Women's Travel Issues*, Washington, D.C.: U.S. Government Printing Office.

Johnson-Anumonwo, I. (1989) 'Journey to work: a comparison of characteristics of single and married parents', *Journal of Specialized Transportation Planning and Practice* 3, 3: 219–46.

Kasarda, J. D. (1988) 'Population and employment change in the U.S.: past, present, and future', paper prepared for the Conference on Long-Range Trends and Requirements for the Nation's Highway and Public Transit Systems, National Research Council, Washington, D.C., June.

Kostyniuk, L. P. *et al.* (1989) 'Mobility of single parents; what do the records show?', *Journal of Specialized Transportation Planning and Practice* 3, 3: 207.

McKeever, C., Quon, J. W. and Valdez, R. (1991) 'Market-based strategies for increasing the use of alternative commute modes', Preprint 91-0291, paper presented to the 70th Annual Meeting of the Transportation Research Board, Washington, D.C., January.

McLafferty, S. and Preston, V. (1991) 'Gender, race, and commuting among service sector workers', *Professional Geographer* 43, 1: 1–15.

Miller, M., Morrison, R. and Vyas, A. (1986) *Minority and Poor Households: Patterns of Travel and Transportation Fuel Use*, Argonne, Illinois: Argonne National Laboratory.

National Center for Health Statistics (1986) 'Aging in the eighties: age 65 and over-use of community services', *Advancedata*, No. 124.

Orski, C. K. (1986) 'Transportation management associations: battling suburban traffic congestion', *Urban Land*, December: 2–5.

Orski, C. K. (1990) 'Can management of transportation demand help solve our growing traffic congestion and air pollution problems?', *Transportation Quarterly* 44, 4: 483–98.

Perez-Cerezo, J. (1986) *Women Commuting to Suburban Employment Sites: An Activity-Based Approach to the Implications of TSM Plans*, Report, Institute of Transportation Studies, University of California, Berkeley, CA.

Pickup, L. (1985) 'Women's travel needs in a period of rising female employment', in G. Jansen *et al.* (eds) *Transportation and Mobility in an Era of Transition*, Amsterdam: Elsevier, North Holland.

Prevedouros, P. and Schofer, J. L. (1991) 'Trip characteristics and travel patterns of suburban residents', paper presented at the 70th Annual Meeting of the Transportation Research Board, National Research Council, Washington, D.C., January.

Rosenbloom, S. (1985a) 'The growth of non-traditional families: a challenge to traditional planning approaches', in G. Jansen *et al.* (eds) *Transportation and Mobility in an Era of Transition*, Amsterdam: Elsevier, North Holland.

Rosenbloom, S. (1985b) *Mothers in the Work Force: The Transportation Implications of the Activity Patterns of Non-Traditional Households in France, The Netherlands, and the United States*, Final Report to the German Marshall Fund of the United States, Austin, Texas.

Rosenbloom, S. (1988a) *The Mobility Needs of the Elderly*, A Report Prepared for the Committee on Improving Mobility and Safety for Older Persons, Austin, Texas, June Version.

Rosenbloom, S. (1988b) 'The mobility needs of the elderly', in Transportation Research Board (ed.) *Transportation in an Aging Society; Improving Mobility and Safety for Older Persons*, Special Report 218, Vol. II, Washington, DC: National Research Council.

Rosenbloom, S. (1989a) 'The impact of growing children on their parents' travel behavior: a comparative analysis', *Transportation Research Record*, 1135.

Rosenbloom, S. (1989b) 'The transportation needs of single salaried mothers: a critical analysis', *Journal of Specialized Transportation Planning and Practice* 3, 3: 255.

Rosenbloom, S. (1991) 'Why working families need a car', in M. Wachs and M. Crawford (eds) *The Car and the City: The Automobile, the Built Environment, and Daily Urban Life*, Ann Arbor, MI: University of Michigan Press.

Rosenbloom, S. with Lerner, A. (1989) *Developing a Comprehensive Service Strategy to Meet Suburban Travel Needs*, Final Report to the Urban Mass Transportation Administration, U.S. Department of Transportation, Austin, Texas.

Rosenbloom, S. and Raux, C. (1985) 'Employment, childcare and travel behavior: France, The Netherlands, and the United States', in *Behavioral Research for Transport Policy*, Utrecht, The Netherlands: VNU Science Press.

Rosenbloom, S. and Remak, R. (1979) *Implementing Packages of Congestion-Reduction Techniques; Strategies for Dealing with the Institutional Problems of Cooperative Programs*, National Cooperative Highway Research Program 205, Washington, D.C.: National Research Council.

Roybal, P. (1991) 'HOV/TSM evaluation study final report', Preprint 91-0115, paper presented to the 70th Annual Meeting of the Transportation Research Board, Washington, D.C., January.

Rutherford, B. M. and Wekerle, G. R. (1989) 'Single parents in the suburbs: journey to work and access to transportation', *Journal of Specialized Transportation Planning and Practice* 3, 3: 277–94.

U.S. Census (1987) *Current Population Reports*, Series P-20, No. 218, Washington, D.C.: Government Printing Office.

U.S. Census (1988) *Statistical Abstract of the United States*, Washington, D.C.: Government Printing Office.

U.S. Department of Transportation (1987) *1983 Personal Travel in the United States*, National Personal Transportation Study, Vol. II, Washington, D.C.: Federal Highway Administration.

Wachs, M. (1978) *Transportation for the Elderly: Changing Lifestyles, Changing Needs*, Berkeley: University of California Press.

Wouters, P. I. J. and Wellman, A. G. (1988) 'Growing old safely', in J. A. Rolhengatter and R. A. de Bruin (eds) *Proceedings of the Second International Conference on Road Safety*, Assen: The Netherlands.

12 WOMEN, THE STATE AND THE LIFE COURSE IN URBAN AUSTRALIA

Allen, J. (1990) 'Does feminism need a theory of "the State"?' in S. Watson (ed.) *Playing the State: Australian Feminist Interventions*, Sydney: Allen and Unwin.

Australian Council on the Ageing and Australian Department of Community Services (1985) *Older People at Home: A Report of a 1981 Joint Survey Conducted in Melbourne and Adelaide*, Canberra: AGPS.

Australian Government (1990) Social Justice Strategy Taskforce Report on Locational Disadvantage, Canberra: unpublished manuscript.

Australian Government (1990–91) *Towards a Fairer Australia: Social Justice Strategy Statement 1990–91*, Canberra: AGPS.

Balbo, L. (1987) 'Crazy quilts: rethinking the welfare state debate from a woman's point of view', in A. S. Sassoon (ed.) *Women and the State*, London: Hutchinson.

Barnett, K. (1988) 'Caring for caregivers: evaluation of the Multicultural Respite Care Project', Adelaide: Multicultural Respite Care Project Advisory Committee.

Barnsley, J. (1988) 'Feminist action, institutional reaction', *Resources for Feminist Research* 17, 3: 18–21.

Borchost, A. and Sim, B. (1987) 'Women and the advanced welfare state – a new kind of patriarchal power', in A. S. Sassoon (ed.) *Women and the State*, London: Hutchinson.

Brennan, D. and O'Donnell, C. (1986) *Caring for Australia's Children*, Sydney: Allen and Unwin.

Burbidge, A. (1990) 'Location of child care in Melbourne', *Family Matters (Journal of the Australian Institute of Family Studies)* 27: 30.

Burke, T. and Hayward, D. (1990) 'Housing Melburnians for the next twenty years: problems, prospects and possibilities', No. 4, Department of Planning and Urban Growth, University of Melbourne, Working Paper.

Coombridge, D. *et al.* (1990) 'Living on a new estate', *Impact Journal of the Australian Council of Social Service* 20, 6: 13–14.

de Lepervanche, M. (1989) 'Breeders for Australia: a national identity for women', *Australian Journal of Social Issues*, 24, 3: 163–82.

D'Mello, A. (1990) 'Child care access: a matter of culture', *Migration Action* 12, 2: 3.

Dollis, N. (1989) 'Access and equity for the ethnic aged', *Migration Action* 11, 3: 14–17.

Fincher, R. (1989) 'Class and gender relations in the local labour market and the local state', in J. Wolch and M. Dear (eds) *The Power of Geography: How Territory Shapes Social Life*, Boston: Unwin Hyman.

Fincher, R. (1991) 'Caring for workers' dependants: gender, class and local state practice in Melbourne', *Political Geography Quarterly* 10: 356–81.

Fincher, R. (forthcoming) 'Child care in Melbourne municipalities: the gender and class relations of access to services in local spaces', unpublished manuscript, Department of Geography, University of Melbourne.

Fincher, R. and McQuillen, J. (1989) 'Women in urban social movements', *Urban Geography* 10: 604–13.

Fincher, R. Webber, M. and Campbell, I. (1989) 'Immigrant women in manufacturing work', unpublished manuscript, Department of Geography, University of Melbourne.

Foy, D. and Crafter, S. (1989) 'Metro planning: social costs and benefits', paper given to the Urban Research Unit, Australian National University, Canberra.

Franzway, S., Court, D. and Connell, B. (1989) *Staking a Claim*, Sydney: Allen and Unwin.

Hernes, H. (1987) 'Women and the welfare state: the transition from private to public dependence', in A. S. Sassoon (ed.) *Women and the State*, London: Hutchinson.

Hess, B. (1985) 'Aging policies and old women: the hidden agenda', in A. S. Rossi (ed.) *Gender and the Life Course*, New York: Aldine.

Hugo, G. (1990) 'Demographic and spatial aspects of immigration', in M. Wooden, R. Holton, G. Hugo, and J. Sloan, *Australian Immigration: A Survey of the Issues*, Canberra: AGPS.

Jonasdottir, A. G. (1988) 'On the concept of interest, women's interests and the limitations of interest theory', in K. B. Jones and A. G. Jonasdottir (eds) *The Political Interests of Gender*, London: Sage.

Jones, K. B. (1988) 'Towards the revision of politics', in K. B. Jones and A. G. Jonasdottir (eds) *The Political Interests of Gender*, London: Sage.

Maas, F. (1990) 'Child care needs of working families in the 1990s', *Family Matters* (*Journal of the Australian Institute of Family Studies*) 26: 59–63.

McCallum, J. and Gelfand, D. (1990) *Ethnic Women in the Middle*, report to the Commonwealth Department of Community Services and Health, Australian National University, National Centre for Epidemiology and Population Health, Canberra.

Marsden, D. and Abrams, S. (1987) '"Liberators", "companions", "intruders" and "cuckoos in the nest": a sociology of caring relationships over the life cycle', in P. Allatt, T. Keil, A. Bryman, and B. Bytheway (eds), *Women and the Life Cycle*, London: Macmillan.

Mayer, K. U. and Muller, W. (1986) 'The state and the structure of the life course' in A. B. Sorenson, F. E. Weinert, and L. R. Sherrod (eds) *Human Development and the Life Course*, Hillsdale, New Jersey: Lawrence Erlbaum Associates.

Monkman, V. (1988) 'Silences: child sexual abuse and the Canadian government', *Resources for Feminist Research* 17, 3: 56–8.

Murphy, P. *et al.* (1990) *Impact of Immigration on Urban Infrastructure*, Canberra: AGPS for the Bureau of Immigration Research.

Pateman, C. (1988) *The Sexual Contract*, Oxford: Polity Press.

Pensabene, T. (1987) 'Multiculturalism and services for the ethnic aged: from philosophy to practice', in C. Foster and H. Kendig (eds) *Who Pays? Financing Services for Older People*, Canberra: ANUTECH.

Perin, C. (1988) *Belonging in America*, Madison: University of Wisconsin Press.

Piven, F. F. (1985) 'Women and the state: ideology, power and the welfare state', in A. S. Rossi (ed.) *Gender and the Life Course*, New York: Aldine.

308

Prentice, S. (1988) 'The "mainstreaming" of daycare', *Resources for Feminist Research* 17, 3: 59–63.

Pringle, R. and Watson, S. (1990) 'Fathers, brothers, mates: the fraternal state in Australia', in S. Watson (ed.) *Playing the State: Feminist Interventions in Australia*, Sydney: Allen and Unwin.

Richards, L. (1990) *Nobody's Home: Dreams and Realities in a New Suburb*, Melbourne: Oxford University Press.

Rossiter, C. (1985) 'Policies for carers', *Australian Journal of Ageing* 4: 3–8.

Rowland, D. T. (1986) 'Immigration and ageing', *Journal of the Australian Population Association* 3: 18–26.

Sassoon, A. S. (ed.) (1987) *Women and the State: the Shifting Boundaries of Public and Private*, London: Hutchinson.

Seitz, A. (1989) 'Non-English speaking background immigrant women: some issues and concerns', *Migration Action* 11, 3: 3–5.

Shaver, S. (1990) 'Gender, social policy regimes and the welfare state', Social Policy Research Centre Discussion Papers No. 26, University of New South Wales.

Victorian Department of Labour (1989) *Child care in Victoria and women's access to the labour force*, Working Paper No. 22, Labour Market Research and Policy Branch.

Watson, S. (1988) *Accommodating Inequality*, Sydney: Allen and Unwin.

Watson, S. (ed.) (1990) *Playing the State: Australian Feminist Interventions*, Sydney: Allen and Unwin.

Yeatman, A. (1990) *Bureaucrats, Femocrats, Technocrats*, Boston: Allen and Unwin.

13 MAKING CONNECTIONS

Beneria, L. and Holdan, M. (1987) *The Crossroads of Class and Gender: Industrial Homeworking, Subcrontracting, and Household Dynamics in Mexico City*, Chicago: University of Chicago Press.

Bowlby, S., Foord, J., McDowell, L. and Townsend, J. (1982) 'Environment, planning and feminist theory: a British perspective', *Environment and Planning A*: 14: 711–16.

Mora, P. (1991) 'stubborn woman', in P. Mora, *Communion*, Houston, TX: Art Publico Press.

Pain, R. (1991) 'Space, sexual violence and social control: integrating geographic and feminist analyses of women's fear of crime', *Progress in Human Geography* 15, 4: 415–31.

Valentine, G. (1989) 'The geography of women's fear', *Area* 21: 85–90.

INDEX

Abrams, S. 259
Adams, K. J. 19
Adamson, N. 206
age stratification 245–6
ageing 157–8
agriculture, *see also* farming, 90, 94–9
Aitken, S. C. 18
Alcraft, R. 136
Allatt, P. 17, 19
Allen, J. 243
Allen, S. 198, 205
analytic ability 99–100, 104–5, 267
Anderson, J. 101
Anstee, M. J. 13
Armstrong, D. 109
Austin, D. 108
Australia 243–63, 275; Home and Community Care 259; Multicultural Respite Care Project 258
autonomy, female 112–14

Back, K. W. 20
Bailey, K. A. 158
Balbo, L. 247
Banting, K. G. 194
Barbados 122–37
Barnett, K. 258
Barnsley, J. 248
Barrett, M. 28
Bartlett, S. 90
Baruch 18
Bateson, M. C. 18–19
beach vendors 136
Beechey, V. 28
Bell, W. J. 287n
Beneria, L. 13, 273
Benston, M. 168
Berger, J. 2

Berleant-Schiller, R. 136
Bernèche, F. 201, 205
Beroldo, S. 239
Besson, J. 122
Bessy, P. 176
Bezirgan, B. Q. 281n
Biel, A. 100
Bird, E. 30, 35, 280n
Bjorklid, P. 101–2, 282n
black lung benefits 164
Blakeley, K. 101–2
Blanc, B. 286n
Blumen, O. 179
Bonvalet, C. 176, 186
Borchorst, A. 206, 247
Bourdieu, P. 203
Bowlby, S. 272
Bracken, D. C. 194
Bradenbaugh, R. 108
Brainn, P. A. 232
Brand, L. E. 204, 286n
Brennan, D. 244
Bridenbaugh, C. 108
Briskin, L. 206
Brodber, E. 108
Bronfenbrenner, U. 101
Brooks 18
Bruner, J. S. 101
Burbidge, A. 255
Burke, T. 256, 261, 263, 288n

Cain, M. T. 93
Campbell, I. 254
Campbell, R. T. 20
Canada 24, 188–207, 248, 270, 271, 275
Canada Assistance Plan 194–5
Carbonara-Moscatti, V. 100

career ambitions 35–6, 44, 58–9
Caribbean 23, 107–31, 122–37, 271, 274
case studies: Alice 49; Alice Roach 117–18; Anne Michado 60–5, 83; Carmen and Pablo 153; Doris 50; Elena 153–4; Etilda and Justo 152–3; Janet Tillman 73–7, 82, 84; Kate 161–2; Lisa Jacobi 65–72, 83; Margery Tuitt Watkins 118–20; Miss Nancy Dyett 114–17; Pat Briggs 77–82, 83–4; Patricia 61; Rita 50; Roberta 51; Rosa and Alirio 152; Sharon 50–1
cash economy 95–6
Castelain-Meunier, C. 172
Chaloupka, C. 288n
Chambers, R. 152
Chaney, E. 5, 16
Chauviré, Y. 176
Chesterfield, R. 93
Chicoine, N. 201, 203–4, 286–7n
childcare 12, 24–5; 66, 112–13, 172; 179; 188–207
child daycare 248, 275; access to 173, 197–8, 203–6; and transport 241; Australia 250–6, 262; and children's age groups 199–200; conceptualization 189–91; cost to parents 196–7; cost-sharing 194; flexibility 202; informal 197–8, 202–4; location 200–2; non-profit cooperative 195, 201; privatization 188; provided by employers 241–2; regulation 191; school services 201–2; separated from education 193–4, 244; state policy 188–207; voluntary 194–5; world patterns 12
childbearing, age of 9, 12
childhood 88
children: access to environment 90; danger to 102–4, 181, 267, 272; excess female mortality 6, 9–10; health and nutrition 145; girls' special experience 22–3, 88–106, 146, 266–7; murdered 102–3; play in industrialized settings 101–3; social policy 244; travel modes 220–3, 226–7, 267; work 89–90, 92–3, 96–7, 112, 117 145–6, 267
Christensen, K. 12–13, 20, 22, 55–87, 269, 273
Christian, P. B. 198, 202

Christopherson, S. 206
Cichocki, M. 55
Clarke, E. 112
Class 4, 24, 142, 154–5, 189, 203, 206, 244, 270, 277; class capacity 24, 26, 249, 251, 253–61, 270, 276
class structure, pioneer settlements 142–3
Cleveland, C. 202
Clifford 28
coal shacks 159–60
coal-mining community 24, 156–70, 271; company power 163–4, 169; mine closure 164, 273–4
cognitive maps 99–100, 104–5
Coleman, M. 101
Collin, J.-P. 195
Colombia 23–4, 138–55, 270–2
community services: access to 245, 253–4; variation in 250–1
commuting time 175, 178–80, 269; telecommuting 67, 86; see also locational decisions
computers and home based work 68
Connell, R. W. 194, 206, 249
Connolly, K. J. 101
Coombridge, D. 255
Coopmans, M. 279n
Corbin, D. A. 159
Cortazar, J. 21
Court, D. 194, 206, 249
Coutras, J. 55
Crafter, S. 255
Craton, M. 109
Croft, S.

Dale, J. 193
Dallacosta, M. 168
dangerous and unsafe environments 102–4, 181, 267, 272
Davidson, D. 240
Davis, K. 28
de Bruin, R. A. 288n
de Lepervanche, M. 244
Deakin, E. 239
Deer, C. D. 150
dependants: elderly 250, 256–61, 262, see also children
Di Gregorio, D. A. 171
Dinnerstein, M. 18
D'Mello, A. 254
Dollis, N. 260
domestic life 108

Douglas, M. 104
Dowling, R. 239
dual-career families 24, 40, 58, 171–87, 194, 270
Dunn, R. 108

earnings *see* wages
economic restructuring, 103
Edelman, M. W. 103
Edholm, R. 168
education 6–9; level of 137; male–female differences 123, 143; separate from child daycare 193–4, 244, *see also* school
Edwards, B. 120
elderly: care of 25, 155, 250–1, 253, 256–61, women 114, 156; and transportation 211, 227–37
Elshtain, J. B. 28
employers: attitude to workers' marital status 37–9; transport initiatives 241–2
employment: downward mobility 48–50, 57; female 113; full-time and part-time 43; and marital status 33–6, 42–52; upward mobility 51, *see also* labour force participation; occupations; work
environmental degradation and children's work 95–6
Ericksen, J. 55
Erikson, E. H. 101
Esping-Anderson, G. 194
ethnicity 4, 12, 88, 189, 202–6, 249, 253, 260, 277; and transportation 211
expectations: changing 60; of life course 56–9; social policy 244

Fagnani, J. 12, 20, 24, 55, 171–87, 267, 270, 275–6
Falk 5
family, postmodern 28–9, 39–41
family organizational pattern: 112; and access to child care 189, 197–8; and locational decisions 180–3, 186
farming: Caribbean 128–35, 137; Colombia 140–3; income from 130, 133, *see also* agriculture
Faure, A. 285n
fear of personal injury 89, 102, 272
Featherman, D. L. 20

Felham, D. 239
female-headed households 33–9, 120, 122–3, 130
feminist movement 172, 248
femocracy 249, 262
Ferguson, E. 239
Fernea, E. W. 281n
Finch, J. 198, 204, 206
Fincher, R. 12, 15–16, 25, 207, 243–63, 275–6, 288n
firms, locational decisions 38–9
Fischloff, B. 104
Flynn, C. P. 239–40
Foster, P. 193
Foy, D. 255
France 24, 171–87, 210, 213, 267, 270, 275
Franzway, S. 194, 206, 249
Frederick, S. J. 239
Fribourg, A. M. 176, 186
Friedan, B. 57
Fuchs, V. 33

Gage, N. 32
Galel el Din, M. 96
Gelfand, D. 256, 259, 262
gender, socially constructed 30, 246–51; division of labour 54, 94, 140, 148, 165, 267
gender relations 41, 151; West Indies 107–21
gendered social policy 246
gendered species 245, 251
geographic knowledge 90
Germain, A. 203, 286n
girls *see* children
Giuliano, G. 239
Glazer, L. J. 239, 240
Godbout, J. 195
Golant, S. M. 279n
Gotman, A. 178
Granger, D. 195
Greenfield, S. 120
Griffith, A. I. 194
Grubb, W. N. 193
Gunn 18
Guyana 124

Hagerstrand, T. 18–19
Handler, J. S. 109
Hanson, P. 210
Hanson, S. 12–13, 20–2, 27–54, 55,

171, 176, 178–9, 187, 207, 210, 269, 275–6, 279n
Hantrais, L. 172
Hareven, T. 19, 20
Harris, L. J. 105
Harris, O. 168
Harrison, M. L. 189
Harrop, A. 279n
Hart, R. A. 18, 90, 100–2, 104, 279n
Hatchuel, G. 188
hawkers 136
Hayward, D. 256, 261, 263, 288n
healthcare insurance and home-based work 84–5
Heilbrun, C. G. 27, 30
Henshall Momsen, J. 13, 16, 23, 122–37, 271, 274
Hermans-Huiskes, M. 279n
Hernes, H. 247
Hess, B. 256
heterosexism 246–7
Higman, B. 109, 120
Hirshey, G. 55
Hochschild, A. 58
home-based 22, 55–87, 269, 273; healthcare 84–5; micro-scale geography of the home 82–4; middle-aged women 72; mothers with young children 60–72; pensions 85; public and private initiatives 85–6
home range 88, 90, 101–2
home–work connections 47, 50, 160–5, 177–80, 194, 240–2
Horvath, F. 60
household composition, shifts in 100, 146, 155
household responsibilty 62–5, 93, 133, 142, 165–6, 172, 179–80, 186; index of 40–1, 47–8, see also work
'housewife' 162; 'housewifisation' 152
houseyards 23, 107–23, 271, 274; described 108–112; see also female life course in 112–14
housing: affordable 263; improvements to 165; ownership 177, 180, 244
housing patterns, Île-de-France 176–7
Howson, J. 109
Holdan, M. 273
Hudson, P. 194
Husband, E. 18

immigrants 158–9, 178–9, 198, 204–5, 207, 253–4, 260, 271, see also migration
immigration policies 198, 244, 270–1, 274–5
industrial segregation 32
Irving, A. 188

Jagdeo, T. P. 112, 114, 120
Jamaica 124
Jean, R. F. 130
job choice 36, 43–7; and location 38–9, 45, 53; sharing 59, 86; skills 48–50, 57, 73–4
Joffe, C. 193
Johansson, S. 10
Johnson 206
Johnson-Anumonwo, I. 179, 210
Jonasdottir, A. G. 249
Jones, K. B. 248
Jopling, C. F. 283n
journey to work 49, 55, 178, 269

Kaniss, P. 55
Kasarda, J. D. 236
Katz, C. 1–28, 88–106, 264–78
Kearney, R. N. 284n
Kellerman, A. 179
Kelly, D. 135
Kenyon, K. L. 239
Kirby, A. M. 104
Kostyniuk, L. P. 210, 288n
Kristof, N. D. 8, 10
Kuiken, J. 195, 286n
Kyle, I. 193

labour force, re-entry 45, 48, 57, 164–5, 269
labour force participation: age-related 124–8, 204; and caring work 256, 272–3, 277; country variations 15; female increase 194; and marital status 33–8; rates 27–54, 256; spatial variations 37–9, 127
labour reserve 141
labour union 163, 284n
Lamphere, L. 41
land-use planning 227, 240–1, 252, 262–3
Lange, F. W. 109
Lazerson, M. 193
Lazreg, M. 281n

Le Grand, J. 191
Le Jeannic, T. 177, 179
Leon de Leal, M. 150
Lerner, A. 227, 239
Lero, D. 193
Lewis, J. 193
Lewis, W. A. 124
Liben, L. S. 105
life course 9, 17; and access to services 25, 199; and access to space 88–9; and change 30, 42, 53, 59, 112; cohort effects 20, 22–3, 42–52, 56, 123, 158, 262, 269; and employment 48, 204; and geographic experience 90, 265; and residential choice 176; as structured by state 199, 243–63; and travel 227–37
life cycle 18–19
life expectancy 5
Lightman, E. 188
locational decisions 171–87; attachment to urban values 183–5; and family organizational pattern 180–3, 186; firms' 38–9; residential 24
locational disadvantage 255–6
Longhurst, R. 93
Lopata, H. Z. 20

Maas, F. 255–6
McCallum, J. 256, 259, 262
McDade, K. 198, 202, 287n
McDowell, L. 207
McIntosh, M. 28
McKeever, C. 240
Mackenzie, S. 194, 286n
McLafferty, S. 211
McLaughlin, S. D. 28
McPhail, M. 206
McQuillen, J. 288n
Madden, J. F. 55, 179
Mann, S. 193
Mariotta, R. 288n
Mark-Lawson, J. 189
marketing 136
Maroney, H. 195
Marsden, D. 259
Martensson, S. 19
Martin, J. 202
Mascarenhas, A. 282n
Massiah, J. 126–7
Massot, A. 176, 285n

maternity leave 69
mating customs: Caribbean 112–14; pioneer settlements 147
Matthews, M. 101
Maurer, W. M. 136
Mayer, K. U. 245
Mayfield, M. I. 285n, 287n
Meertens, D. 143, 148
Meintel, D. 198
Mesa-Lago, C. 16
Meyer, J. 279n
Michelson, W. 171, 182
Mies, M. 94, 151
migration 111, 119, 147–8; female 113, 127, 137–8, 148, 182, 268, 270–2; male, outward 95, 97–8, 99, 123, 268; rural-to-rural 23–4, 138–55, 268; seasonal 140, 148; selective 148–9
Miller, B. D. 284n
Miller, M. 211
mobility 265–72
Molano, A. 151
Monk, J. 1–28, 89, 264–78, 279n
Monkman, V. 248
Montserrat 111, 122–37
mothers/mothering 9, 24–5, 53, 58, 60, 66, 181, 209, 246–7, 268; 'good mother' 58–9, 172, 181, 270; lone mother 194, 197–8, 204, 210
Moore, M. 194
Mora, P. 264
Morris, L. 286n
Morrison, R. 211
Moser, C. O. N. 154
Moss, P. 188, 202, 206
Mourant, R. R. 288n
Muller, W. 245
Munroe, R. H. 99–100, 104
Munroe, R. L. 99–100, 104
Murphy, P. 288n
Myles, J. 206

Nag, M. 93
neighbourly relations 164, 274; exchange 166; neighbouring 157, 166–7
Neitzert, F. 178
Nelson, K. 37, 38
Nerlove, S. B. 99–100, 104
Netherlands 208, 210, 212, 213, 223–7, 237, 267

Neugarten, B. 19
Nevis 122–37
Nigeria 93
Nygren, D. 10

occupations: age-related 128–35, 135–6; segregation 29–30, 32
O'Connor, J. S. 194
O'Donnel, L. 179
O'Donnell, C. 244
off-farm jobs 128–30, 133–5, 137, 154–5
old age 147–8
Olson, A. 12, 13
Olwig, K. F. 109
Orski, C. K. 239–40
Ouellet, A. 285n, 286n
Ozawa, M. N. 20

Pain, R. 105, 272
Palm, R. 55
Parr, J. 52
Pascall, G. 193
Pateman, C. 246
Patterson, O. 120
Peet, R. C. 93
Pence, A. R. 193
Pensabene, T. 260
pensions 85, 137, 147, 164
Perez-Cerezo, J. 210
Perin, C. 252
Peterson, B. R. 28
Phillips, A. 28, 188, 202, 206, 279n
Pickup, L. 210
Pinçon, M. 184
Pinçon-Charlot, M. 184
pioneer settlements 138–55; class structure 142–3; economic organization 140–2; gender roles in 139–40
Piven, C. 248
Pollowy 282n
power relations 94, 151
Pratt, G. 12–13, 20–2, 27–55, 171, 176, 178, 179, 187, 207, 269, 275–6
Pred, A. 18, 55
Prentice, S. 191, 194, 196, 248, 286n
Preston, V. 211
Prevedouros, P. 210
Pringle, R. 33, 37, 247, 279n
production, capitalist relations 95; and reproduction 3, 12, 21, 23–5, 94–5, 142, 145, 168–9, 172, 181, 207, 265, 272–7
productivity, rural 99
public space 89, 272
public and private spheres 172, 186, 272
Pulsipher, L. M. 12, 16–17, 23, 107–21, 266–7, 271, 274, 283n
purdah 88–9, 93–5, 98, 103, 106, 270

Quint, M. 288n
Quon, J. W. 240

race 4, 12, 88; and transportation 211
Rackoff, N. J. 288n
Raux, C. 210
Redclift, N. 28
Reddock, R. 124
Remak, R. 239
remittances 113, 128, 133, 137, 271
Rendu, P. 184
residential decision 24
residential mobility 171–87; methodology 173–5; reasons for moving 177–8
Reskin, B. 29
respite care 258–9
retirement 165, 169
Rhode Island 41
Richards, L. 255
risk perception 104
Risser, R. 288n
Roberts, L. J. 283n
Robins, B. 55
Robinson, B. E. 101
Robinson, R. 191
Rodgers, G. 93
Rolhengatter, J. A. 288n
rootedness 271–2
Rose, D. 12, 24, 179, 188–207, 254, 270–1, 274–5, 282n, 286–7n
Rosenbloom, S. 12, 15, 16, 25, 208–42, 267–9, 271, 275, 287n
Rosenzweig, R. 32
Rosser, S. V. 17
Rossi, A. 19, 20, 171
Rossiter, C. 258
Rowland, B. H. 101
Rowland, D. T. 260
Rowles, G. D. 18, 279n
Roybal, P. 239
Ruddle, K. 93

Ruggie, M. 193
Rutherford, B. M. 211

Saarinen, T. 18
Sachs, P. 15–17, 24, 156–70, 271, 273, 274
Saegert, S. 101, 182
St Vincent 122–37
San Francisco 38
Saraceno 20
Sassoon, A. S. 207, 250
Savage, M. 189
scale, and the life course 265, 272–6
Schildkrout, E 93
Schofer, J. L. 210
school 146; attendance 98–9; and migration 149; and spatial range 146; childcare services 201–2; enrolment 6–8, 96, 127, *see also* education
Seager, J. 12–13
Seagert 104
Secombe, W. 168
Seitz, A. 260
Self, C. M. 105
Selinger, G. 195
Sell, J. L. 18
sequential shift strategy 41
Seward, S. B. 198, 202, 287n
Shaver, S. 246–7
Shlay, A. 171
Siim, B. 206
Siltanen, J. 29, 35, 40
Sim, B. 247
Sivard, R. L. 4, 13
Skelton, T. 135
Smith, D. E. 194
Smith, J. 29
Smith, M. G. 120
Smith, P. 104
Smith, R. T. 110, 114, 120, 283n
social control, and spatial forms 88
social policy 243–63
social reproduction 168–9; practices 93–4, 247; work of 142, 146–8, 150, 207, 269, 272–3
social services, country variations 16, *see also* community services
Soja, E. W. 2
Sorrentino, C. 193, 286n, 287n
space: access to and control of 88; gendered 251–2, 272
spatial competence 267

spatial policy 262–3
spatial range, children 88–9, 146, 150, 266–7, 272
spatial relations between home and work 39, 45, 53
spatial variations in employment 41
spatio-temporal aspects of women's work 12, 44, 186, 207, 240–1
Spence, E. J. 136
Stacey, J. 28–9, 39, 54
Standing, G. 93
Stapleton, C. 18, 171, 187
State, the 243–63; 'constructs' women 246–7; instruments of action 245–6; oppresses women 247–8; services 12, 188–92, 206
Stefani, R. L. 283n
Stewart, J. M. 120
Stueve, A. 179
suburban loneliness 25–6, 241, 255–6
Sudan 22–3, 88–106, 266–7, 270; Gezira Scheme 96; Suki Agricultural Development Project 90, 94–9
Sweden 210, 241
Swedenburg, T. 281n

Taffin, C. 176
Taylor, P. 204
tenancies, Sudan 97
Thompson, L. 172, 285n
Tilley, L. A. 30
Tindal, M. A. 90, 101–2
Torell, G. 100
Townsend, J. G. 6, 13, 16–17, 23–4, 138–55, 267–8, 270, 272, 274
transport 25; and childcare 202; employer initiatives 241–2; mandatory transportation demand (TDM) programmes 239–40; public 236, 238–9, 241, 255; subsidized 209, 237
travel modes, children's 220–3, 226–7, 270
travel patterns 208–42, 275–6; elderly women 25, 209, 211, 227–37, 238, 271; gender differences 214–220; linked trips 214–16, 276; the literature 210–12; and marital status 210–11, 268, 213–27, 229–30, 237–8; parents' 210–11, 213–27, 237–8, 240–1; racial and ethnic variables 211; serve-passenger trips 216–220, 223–6, 269

Treiman, D. J. 30, 280n
Truelove, M. 194, 286n
Turner, V. W. 283n

United Kingdom 210, 241
United States 22, 28–9, 252, 266–9,
    271, 273, 275; children in cities
    88–106; Deckers Creek 156–70;
    home-based work 55–87; Small
    Business Administration (SBA)
    75, 86; women's travel pattern
    208–42; Worcester, Massachusetts
    31–52
urban consolidation 262
urban form 171, 176; and gender
    240–2, 244–5

Valdez, R. 240
Valentine, G. 105, 272
Villeneuve, P. 179
Vyas, A. 211

Wachs, M. 211
Waerness, K. 206
wages: component wage 29, 35; family
    54, 247, 273; low 27–8, 35–6
Walker, A. J. 172, 285n
Wallman, S. 200
Warde, A. 189
Warshaw, M. 32
Wash, D. P. 204, 286n
water, piped 98
Watson, S. 244, 247, 249
wealth distribution, geographic 6
Webber, M. 254
Weiss, R. S. 285n
Wekerle, G. R. 211
welfare state institutions 248, *see also*
    community services
Wellmann, A. G. 235
Wells-Bowie, L. 283n
Wentworth, T. 109, 120
West, J. 30, 35, 280n
West Indies 107–21, 122–37
Westmacott, R. 283n
White, B. N. F. 93

White, J. W. 55
White, R. 90
White, S. 279n
Wildavsky, A. 104
Willis, T. 202
Wilson de Acosta, S. 145, 148
women: conflicting obligations 40, 43,
    172–3, 179; as dependants 247;
    elderly 25–6, 147–8, 156, 169, 209,
    mobility 15–16, 25, 227–38, 270–2;
    geographic distribution, by age 5, of
    older women 13–15; life-course
    literature 17–21; middle-aged
    42–52, 72, 256–61, 275; world
    patterns 5–17; with young children
    60–72
women's movement 57–8, 248–9, 270,
    277
work: agricultural 94–5, 149–52;
    children's 89–90, 92–3, 96–7, 130,
    145–6, 267; domestic 40–1, 53, 58,
    63–4, 159–63, 165–6, 168, 223;
    hours of 133–5; job choice 36, 38–9,
    43–7, 53; journey to 178–80, 269;
    meaning of 52, 168; off-farm jobs
    128–30, 133–5, 137; part-time or
    temporary 46, 55, 73, 86;
    production/reproduction link
    148–152, 154, 168–9, 208, 237, 247,
    256, 272–7; 'volitional' 169, *see also*
    employment; home-based work;
    household responsibility; labour
    force participation; occupation
work force: female reserve 135; by
    gender and age 143–8; first time
    entry 164–5; gender roles 13, 20,
    149–50; older women 15, 22; return
    to 45–6, 48, 57
Wouters, P. I. J. 235
Wycko, W. 239

Yeandle, S. 207
Yeatman, A. 249
youth 88

Zelinsky, W. 279n